SPARKNOTES™
101

Algebra

SPARK PUBLISHING

Written by Anna Medvedovsky.
Expert reviewed by Daniel Ketover.

Spark Publishing
A Division of Barnes & Noble Publishing
120 Fifth Avenue
New York, NY 10011
www.sparknotes.com

Please submit all changes or report errors to www.sparknotes.com/errors.

Printed and bound in the United States.

Library of Congress Cataloging-in-Publication Data

SparkNotes 101 : algebra.
    p. cm.
    Includes index.
    ISBN-13: 978-1-4114-0334-5
    ISBN-10: 1-4114-0334-7
  1. Algebra—Problems, exercises, etc. 2. Universities and Colleges—
Entrance examinations—Study guides. I. Title. II. Title: Algebra.

QA157.M43 2006
512—dc22

                                        2006010322

Contents

A Note from SparkNotes

Welcome to the *SparkNotes 101* series! This book will help you succeed in your introductory college course for algebra.

Algebra is everywhere. Whether you're following a recipe or balancing your checkbook, having a solid foundation in algebra will not only help you navigate the world around you but also help you understand the more complex and conceptual mathematics that you'll encounter in more advanced courses.

Every component of this study guide has been designed to help you process the course material more quickly and score higher on your exams. The format of this book will allow you to easily situate yourself and get to the crux of your course. We've organized the book in the following manner:

Chapters 1–9: Each chapter provides a clarification of material included in your textbook. Key features include:

- **Examples:** Plenty of examples are worked out in step-by-step detail, simulating the tutoring experience. Whenever you can, try to solve these examples on your own before reading the solutions.

- **Key Points:** Throughout the text, these boxes will highlight quick reviews, formulas, and important points.

- **Tricky Points:** Some math concepts can be particularly complex. These boxes help clarify and simplify some of the more thorny ideas.

- **Options:** Wherever possible, we'll show you different ways to solve problems so that you can decide which method works best for you.

- **Summary:** These end-of-chapter recaps provide at-a-glance reviews of major topics.

- **Sample Test Questions:** These show you the kinds of questions you are most likely to encounter on a test. Answers are provided at the end of the book.

SparkChart: Look for the fold-out chart in the middle of the book to make studying for exams quick and easy.

Glossary: The math terminology used in this book is defined and redefined within the text whenever it comes up. But if you need a little help with an unfamiliar word, check the glossary at the end of the book.

We hope *SparkNotes 101: Algebra* helps you, gives you confidence, and occasionally saves your butt! Your input makes us better. Let us know what you think or how we can improve this book at **www.sparknotes.com/comments.**

Equations and Solutions

1

Overview

Say little Ethel has four more apples than little Thelma. If Ethel has nine apples, how many does Thelma have?

Okay, you may think, I can do this in my head. If Ethel's nine apples are four more than Thelma's number, Thelma must have 9 – 4 = 5 apples. Not so tough.

Try another. Ethel and Thelma are carrying a bag of apples home to their mother. On the way, Thelma eats half of all the apples. Ethel finds seven more, but Thelma eats a third of the new total amount. When they get home, they find that they have two more apples than they started with. How many apples were in their bag in the beginning?

This problem has so many details that it's difficult to keep track of all the pieces of information. Algebra is one popular way of making such a problem manageable: the information in the problem is used to construct an *equation* whose *solution* gives you the answer.

Expressions

ARITHMETIC EXPRESSIONS

An **arithmetic expression**, or simply *expression*, is one or more numbers combined together using only the four basic operations (addition, subtraction, multiplication, and division), possibly involving parentheses. All of the following are arithmetic expressions:

$$-7.5 \qquad 10 + 3 \qquad 8 \cdot \frac{3}{2} \qquad \left(4 - \frac{7}{8}\right) \div 2$$

An arithmetic expression can always be **evaluated**, or simplified, to a real number. If the expression involves more than one operation, we perform each operation in the conventional **order of operations**:

KEY POINTS

Order of Operations

1. Evaluate expressions in **P**arentheses. Follow order of operations.

2. Evaluate **E**xponential expressions.

3. Do **M**ultiplication and **D**ivision, from left to right.

4. Do **A**ddition and **S**ubtraction, from left to right.

Common mnemonic: **PEMDAS**

EXAMPLE: Order of operations

Evaluate the expression $8 - 35 \div ((23 - (8 - 7)4^2))$.

SOLUTION

Use order of operations to determine that the operations must happen in the following order:

$$8 \overset{6}{-} 35 \overset{5}{\div} \left(23 \overset{4}{-} (8 \overset{1}{-} 7) \overset{3}{\times} \overset{\overset{2}{\frown}}{4^2} \right)$$

So,

$$
\begin{aligned}
8 - 35 \div (23 - (8 - 7) \times 4^2) &= 8 - 35 \div (23 - 1 \times 4^2) & \text{Step 1} \\
&= 8 - 35 \div (23 - 1 \times 16) & \text{Step 2} \\
&= 8 - 35 \div (23 - 16) & \text{Step 3} \\
&= 8 - 35 \div 7 & \text{Step 4} \\
&= 8 - 5 & \text{Step 5} \\
&= 3 & \text{Step 6}
\end{aligned}
$$

So the expression $8 - 35 \div (23 - (8 - 7)4^2)$ evaluates to 3.

Variables

In algebra, quantities are often unknown or unspecified. Such quantities are called **variables**. A variable is usually designated by a letter of the alphabet, like x, y, a, b, or p.

For example, let the letter a represent the number of apples that Ethel has. If she eats 2 apples, she will have $a - 2$ apples. If Thelma then gives her 4 apples, Ethel will have $a - 2 + 4$ apples. The variable a allows us to keep track of Ethel's apples.

There are some conventions for which letters get used for variables in which contexts, but ultimately it doesn't matter very much as long as you're consistent.

Algebraic Expressions

An **algebraic expression** is any combination of variables and numbers using the four basic operations:

$$a - 2 \qquad 17 + x \qquad \frac{b}{3} \qquad \frac{c}{d} + 2 \qquad 13 - (2)(8)$$

An algebraic expression need not involve variables—so every arithmetic expression is also an algebraic expression. For example, $13 - (2)(8)$ above is both an algebraic and an arithmetic expression. We'll use the word *expression* to mean any kind of algebraic expression, with or without variables.

In this section, we'll use variables and expressions to keep track of specific yet unknown quantities. For example, $a - 2 + 4$ is an algebraic expression that depends on a, the variable. We say that $a - 2 + 4$ is an algebraic expression **in terms of** a.

Constants

A **constant** is a quantity that doesn't change. All real numbers are constants; their value stays the same. A popular real-number constant may be represented by a letter. For example, the Greek letter π is a constant that represents, by convention, the real number 3.14159 . . . , the ratio of any circle's circumference to its diameter.

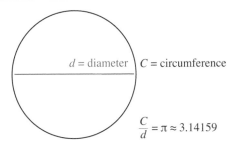

d = diameter C = circumference

$$\frac{C}{d} = \pi \approx 3.14159$$

The word *constant* is also used to designate the real numbers in algebraic expressions. In $2a - 5b + 3$, the 3 is the constant term.

Physicists have to work with many difficult-to-express numbers, so they often use constants for convenience. The letter g often stands for 9.8 meters per second squared, the acceleration due to gravity.

While π and 3 and g are examples of constants that never change their value, other constants depend on context. For example, let's say you have a calculator that has only one button: multiplication by b. So if you enter any number x into the calculator, it will spit out bx. In this example, b is a constant (although it looks like a variable) and x is a variable.

EVALUATING EXPRESSIONS

An expression without a variable, like $6(5 - 2)$, can always be *evaluated* to a unique real number. An expression that involves a variable, like $a + 7$, takes on different values depending on the value of the variable. For example, when $a = 8$, the expression $a + 7$ evaluates to $8 + 7 = 15$. When $a = -3$, the same expression evaluates to $-3 + 7 = 4$.

Evaluating an expression for a particular value of the variable is colloquially called **plugging in** a value, or **substituting** for the variable. If you plug the value $x = 9$ into the expression $5x - 2$, you get $5(9) - 2 = 43$.

Be sure to plug in the same value each time the variable appears in the expression. To evaluate $y + 5 - 3y^2$ when $y = -2$, replace both ys in the expression with -2s:

$$y \quad + 5 - 3 \ y^2$$
$$\downarrow \qquad\qquad \downarrow$$
$$(-2) + 5 - 3(-2)^2$$

This new expression now has no variables and can be evaluated using order of operations:

$$(-2) + 5 - 3(-2)^2 = (-2) + 5 - 3(4) = (-2) + 5 - 12 = -9$$

Be careful about matching values and variables when there is more than one variable.

EXAMPLE: Plugging in two variables

Evaluate the expression $(a + b)a$ when $a = 2$ and $b = 3$.

SOLUTION

Rewrite the expression and plug in:

$$(a + b) \ a$$
$$\downarrow \ \downarrow \ \downarrow$$
$$(2 + 3) \ 2 = 10$$

SIMPLIFYING EXPRESSIONS

An expression that involves several operations can often be **simplified** to a **simplified form**. Let's see how this actually works.

EXAMPLE: Adding constant terms
Simplify $a + 7 - 2$.

SOLUTION
No matter the value of a, adding 7 and subtracting 2 is the same thing as adding 5. Watch how this works on the number line:

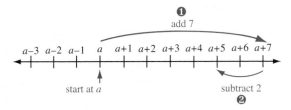

The essential idea is that you can work with the constants and the variable separately: $a + 7 - 2$ is the same thing as $a + (7 - 2)$, which is $a + 5$.

EXAMPLE: Adding variable terms I
Simplify $3b - b$.

SOLUTION
You can just think this one through as if b were an apple or banana: if you have three bs and you take away one b, you're left with two bs: $3b - b = 2b$. In symbols (or **algebraically**), $3b$ is the same thing as $b + b + b$, so

$$3b - b = (b + b + b) - b = b + b = 2b$$

EXAMPLE: Adding variable terms II

Simplify $4 \cdot 9z$.

SOLUTION

Again, you can think this one through: four times nine zs gives thirty-six zs.

Algebraically, the idea is that you can think about multiplying the constants and the variable separately:

$$4 \cdot 9z = (4 \cdot 9)z = 36z$$

The Distributive Property

Simplifying and evaluating expressions depends on various properties of real numbers. We're so used to seeing most of these properties in action that it doesn't seem particularly illuminating to trot them out and give them each a name. For example, the fact that $3 + 4$ is the same thing as $4 + 3$ happens to be called the **commutative property of addition**.

However, there is one property worth noticing. It's called the **distributive property**, and it's more exciting than the rest because it tells us how addition and multiplication are related. You'll also find this property very handy for simplifying expressions.

> *Distributive Property*
>
> If a, b, and c are real numbers, then
>
> $$a(b + c) = ab + ac$$

We can also do the same thing with subtraction or switch the order of a and $b + c$, so that we get four statements of the distributive property:

> *Distributive Property*
>
> If a, b, and c are real numbers, then
>
> $$a(b + c) = ab + ac$$
> $$a(b - c) = ab - ac$$
> $$(b + c)a = bc + ca$$
> $$(b - c)a = ba - ca$$

Now try some examples that use the distributive property.

EXAMPLE: Distributing with a plus sign

Simplify $2(c + 7)$.

SOLUTION

The distributive property tells us how to deal with this one.

$$2(c + 7) = 2 \cdot c + 2 \cdot 7$$
$$= 2c + 14$$

EXAMPLE: Distributing with a minus sign

Simplify $3(8 - 2y)$.

SOLUTION

Again, the distributive property is the way to go. Make sure you distribute the 3 to both terms in parentheses.

$$3(8 - 2y) = 3 \cdot 8 - 3 \cdot 2y$$
$$= 24 - (3 \cdot 2)y$$
$$= 24 - 6y$$

Now watch how some of these examples combine:

EXAMPLE: Combine like terms

Simplify $4d + 2 - \frac{1}{2}d - 5$.

SOLUTION

We can rearrange the order of the *terms* (the parts being added):

$$4d + 2 - \frac{1}{2}d - 5 = 4d - \frac{1}{2}d + 2 - 5$$

Now we can group the variable terms together and the constant terms together:

$$4d - \frac{1}{2}d + 2 - 5 = \left(4d - \frac{1}{2}d\right) + (2 - 5)$$

To figure out $4d - \frac{1}{2}d$, you can think it through again: four *d*s minus half a *d* gives three and a half *d*s. Algebraically, it's the distributive property backward that lets us think of $4d - \frac{1}{2}d$ as $\left(4 - \frac{1}{2}\right)d$, which simplifies to $\left(3\frac{1}{2}\right)d$, or $\frac{7}{2}d$.

All together, the solution should look like this:

$$4d + 2 - \frac{1}{2}d - 5 = 4d - \frac{1}{2}d + 2 - 5$$

$$= \left(4d - \frac{1}{2}d\right) + (2 - 5)$$

$$= \left(4 - \frac{1}{2}\right)d + (-3)$$

$$= \frac{7}{2}d - 3$$

This next example is similar; watch those pesky minus signs.

EXAMPLE: Distribute and combine

Simplify $9 - 2(3 - p)$.

SOLUTION

$$
\begin{aligned}
9 - 2(3 - p) &= 9 - 2 \cdot 3 - 2 \cdot (-p) \\
&= 9 - 6 - (-2p) \\
&= (9 - 6) + 2p \\
&= 3 + 2p
\end{aligned}
$$

The following example puts everything together. Try it on your own before looking at the solution.

EXAMPLE: Harder: distribute and combine

Simplify $-\dfrac{1}{2}(-2 + 8s) + 4(3s - 1)$.

SOLUTION

$$
\begin{aligned}
-\frac{1}{2}(-2 + 8s) + 4(3s - 1) &= -\frac{1}{2} \cdot (-2) - \frac{1}{2} \cdot (8s) + 4 \cdot (3s) + 4 \cdot -1 \\
&= \frac{1}{2} \cdot 2 - \left(\frac{1}{2} \cdot 8\right)s + (4 \cdot 3)s - 4 \\
&= 1 - 4s + 12s - 4 \\
&= (-4s + 12s) + (1 - 4) \\
&= (-4 + 12)s + (-3) \\
&= 8s - 3
\end{aligned}
$$

The last example ups the ante a bit: there are *two* variables involved.

EXAMPLE: Two-variable distribute and combine

Simplify $3(a + b + 1) - 2(b + 4)$.

SOLUTION

First, distribute the 3 across the first set of parentheses:

$$3(a + b + 1) - 2(b + 4) = 3 \cdot a + 3 \cdot b + 3 \cdot 1 - 2(b + 4)$$
$$= 3a + 3b + 3 - 2(b + 4)$$

Next, distribute the 2 to the two terms enclosed by the remaining parentheses:

$$3a + 3b + 3 - 2(b + 4) = 3a + 3b + 3 - 2 \cdot b - 2 \cdot 4$$
$$= 3a + 3b + 3 - 2b - 8$$

To simplify completely, we have to combine **like terms**—those terms that involve the exact same variables (to the same power). In this example, 3 and –8 are like terms (both are constants), and 3*b* and –2*b* are like terms (both are *b*s). So we can group them together and then add them:

$$3a + 3b + 3 - 2b - 8 = 3a + (3b - 2b) + (3 - 8)$$
$$= 3a + b - 5$$

The expression $3a + b - 5b$ has no uncombined like terms; it is completely simplified.

After all of these examples, we're ready to describe the features of a *simplified* expression: it involves no parentheses and no uncombined like terms. Most important, a simplified expression is completely **equivalent** to the original expression, which means that you can get from one to the other using the properties of real numbers and operations.

KEY POINTS

Simplifying Expressions

1. Get rid of all parentheses: use the distributive property.

2. Combine like terms: use the properties of addition to group like terms together.

3. Use order of operations to evaluate all number expressions.

SIMPLIFYING AND EVALUATING EXPRESSIONS

If you're asked to evaluate an expression, should you simplify before plugging in? For example, if you're asked to evaluate $4\left(\frac{3}{2} - x\right) + x$ when $x = -3$, should you first clear out the parentheses or plug in –3 for x?

It all depends—on you. You should get the same answer either way.

- If you're good with numbers but shaky with variables, plug in the values for the variables early on. Then simplify the variable-free expression using brute force.

- If you're likely to forget a minus sign or make a multiplication mistake and variables don't make you nervous, simplify first. Then plug in the values for the variables at the end, once the expression is simplified. Some people, including many teachers, prefer to simplify first because it's cleaner and more elegant.

Neither method is more correct, and that's exactly what makes variables such a great tool.

To demonstrate, let's go through that example, both ways.

EXAMPLE: Option one: plug in first

Find the value of $4\left(\frac{3}{2} - x\right) + x$ for $x = -3$.

SOLUTION

It often helps to keep the value you're substituting in parentheses at first, especially when that value is negative, as it is here. You're less likely to drop a negative sign if you write everything out step by step.

$$4\left(\tfrac{3}{2} - x\right) + \quad x$$

$$\text{Substitute } x = -3 \qquad \downarrow \qquad \downarrow$$

$$4\left(\tfrac{3}{2} - (-3)\right) + (-3) = 4\left(\tfrac{3}{2} + 3\right) + (-3)$$

$$= 4 \cdot \tfrac{3}{2} + 4 \cdot 3 - 3$$

$$= 6 + 12 - 3$$

$$= 15$$

EXAMPLE: Option two: simplify first

Find the value of $4\left(\dfrac{3}{2} - x\right) + x$ for $x = -3$.

SOLUTION

To simplify the expression, distribute the 4 to clear out the parentheses and then collect like terms:

$$4\left(\frac{3}{2} - x\right) + x = 4 \cdot \frac{3}{2} - 4 \cdot x + x$$
$$= 6 - 4x + x$$
$$= 6 - 3x$$

Finally, plug $x = -3$ into $6 - 3x$ to get $6 - 3(-3) = 6 + 9 = 15$.

We got the same answer both times, so we know we did it right. If we had gotten two different answers, we would have known that we made a mistake.

OPTIONS

Choose Your Method Wisely Whether you prefer to simplify first or to plug in first, there are some situations when one method is clearly preferable. If one way proves difficult, start over with the other.

Take a look at two extreme examples.

EXAMPLE: Easier to simplify first

Evaluate $2(x + 4) - 2x$ when $x = 1567$.

SOLUTION

Simplifying first, you'll get $2(x + 4) - 2x = 2x + 8 - 2x = 8$: the variables cancel. So the value of the expression is 8 regardless of the value of x! Compare this to working out a monster like $2(1567 + 4) - 2(1567)$ if you decide to plug in first.

Moral: If the variables cancel, simplifying first is easier. Also, if the value you're plugging in is unsightly, simplifying first is easier.

EXAMPLE: Easier to plug in first

Find the value of $\frac{1}{4}\left(z - \frac{5}{3}\right)$ for $z = \frac{17}{3}$.

SOLUTION

The fact is that $z - \frac{5}{3} = 4$ makes the rest of the calculations easy—if you plug in first: $\frac{1}{4}\left(\frac{17}{3} - \frac{5}{3}\right) = \frac{1}{4} \cdot \frac{12}{3} = 1$. But if you simplify first, you're in for some unpleasantness:

$$\frac{1}{4}z - \frac{1}{4} \cdot \frac{5}{3} = \frac{1}{4}z - \frac{5}{12}$$

Moral: If the value of the variable helps clean up the expression, plugging in first may be easier.

Equations

An **equation** is a statement that says that two expressions have the same value. An equation always involves an equal sign; an expression never does. For example, $2 + 4 = 6$ is an equation: it says that the expression $2 + 4$ and the expression 6 represent the same value.

An equation may involve variables; sometimes we specify how many. For example, $x + 1 = 9$ is an *equation in one variable*; $2a + b = c + 7$ is an *equation in three variables*.

TRUE AND FALSE EQUATIONS

An equation may be true or it may be false, the same way that English-language statements may be true or false. For example, the equation $1 + 3 = 4$ is true, but the equation $2 + 2 = 5$ is false. It's important to be able to recognize false equations because they may point to false assumptions or false reasoning.

Sometimes we don't have enough information to determine whether an equation is true or false. For example, the equation $x + 1 = 4$ on its own is neither true nor false—it all depends on the value of x. If the value of x is not known or not specified, we can't say anything about whether $x + 1 = 4$ is true. This equation happens to be true when $x = 3$ and false for other values of x.

TRICKY POINTS

Always-True and Always-False Equations Sometimes we can determine whether an equation is true or false even if we don't know the value of the variables.

Always true: The equation $x + x = 2x$ is true regardless of the value of x: no matter what value of x you plug in, $x + x = 2x$ is a true statement. So we can say that $x + x = 2x$ is always true. Any value of x works to make it a true statement.

Always false: On the other hand, the equation $y + 1 = y$ is false regardless of the value of y: no matter what value of y you plug in, $y + 1 = y$ is a false statement. So we can say that $y + 1 = y$ is always false. No value of y will make it into a true statement.

SOLUTIONS TO AN EQUATION

A **solution** to an equation is a value for the variable that makes that equation true. For example, $x = 2$ is a solution to the equation $5 - x = 3$ because when you plug in 2 for x, you get a true equation: $5 - 2 = 3$.

Equations can have different types of solutions:

- **A solution may be a set of values:** A solution to a one-variable equation is just a number, as in the example above. A solution to a multi-variable equation is a *set of numbers*.

  For example, one solution to the equation $a = b + 4$ is $a = 4$, $b = 0$. Check that this is true by plugging in for the variables: $4 = 0 + 4$ is a true statement.

- **A solution may not exist:** An equation may or may not have solutions.

  As mentioned earlier, there is no value of y that makes the equation $y + 1 = y$ true. This equation has no solutions.

- **An equation may have many solutions:** A solution, if it exists, may not be unique. The equation $a = b + 4$ has more than one solution. For example $a = 5$, $b = 1$ works just as well as $a = 4$, $b = 0$.

Finding solutions to equations is the name of the game in algebra—we'll be spending much of the rest of the book on different methods of **solving** different kinds of equations.

Checking Solutions

To check whether a particular real number is a solution to a single-variable equation, simply plug in the number for the variable wherever it occurs, on both sides if necessary, and see if you end up with a true statement.

EXAMPLE: Check solutions in linear equation

Is –4 a solution for $3x + 45 = 13 - 5x$?

SOLUTION

Plug in and see!

$$3x + 45 = 13 - 5x$$
$$3(-4) + 45 \stackrel{?}{=} 13 - 5(-4)$$
$$-12 + 45 \stackrel{?}{=} 13 - (-20)$$
$$33 = 33$$

Since we ended up with a true statement, 33 = 33, we know that –4 is indeed a solution to the equation $3x + 45 = 13 - 5x$.

EXAMPLE: Check solutions in quadratic equation

Check whether 1 and 2 are solutions to the equation $(a + 2)(a - 3) = -4$.

SOLUTION

Again, plug in and see:

$$(a + 2)(a - 3) = -4$$

Plug in $a = 1$ | Plug in $a = 2$

$$(1 + 2)(1 - 3) \stackrel{?}{=} -4 \qquad (2 + 2)(2 - 3) \stackrel{?}{=} -4$$
$$(3)(-2) \stackrel{?}{=} -4 \qquad\qquad (4)(-1) \stackrel{?}{=} -4$$
$$-6 \neq -4 \qquad\qquad\qquad -4 = -4$$

Plugging in $a = 1$ simplifies to the false statement $-6 = -4$, so 1 is not a solution to $(a + 2)(a - 3) = -4$. But plugging in $a = 2$ gives the true statement –4 = –4, so 2 *is* a solution to the equation.

In the next section we'll learn how to actually find solutions to equations such as $3x + 45 = 13 - 5x$; we'll return to equations like $(a + 2)(a - 3) = -4$ in a few chapters. Regardless of the methods we use or the complexity of the equation, checking your solutions is always a good idea. It only takes a few seconds, and it gives unparalleled peace of mind.

Solving Equations

In this section, we'll go over how to solve equations that, after each side is simplified, look like $Ax + B = Cx + D$. These equations are called **linear** equations in one variable. They're called *linear* because they're related to equations whose graphs are straight lines. We'll discuss this more in the next chapter.

The method of solution is always the same: we take an equation and manipulate it to give an **equivalent** equation whose solution is obvious. Two equations are *equivalent* if they have the same solutions.

What's an example of an equation whose solution is obvious? Here's one.

EXAMPLE: Finding solutions to an equation
Find all the solutions to the equation $x = -1$.

SOLUTION
Only $x = -1$ will make this equation true.

SOLVING BY ADDING

What are the solutions to $x + 7 = 9$? We're looking for a number so that when we add 7 to it, we'll get 9. The only value of x that works is $9 - 7$, or 2.

The additive property of equality expresses this same idea in mathematical terms.

> ### Additive Property of Equality
> For any real numbers a, b, and c, the equation
>
> $$a = b$$
>
> is equivalent to the equation
>
> $$a + c = b + c$$

In other words, you can add the same number to both sides of an equation. And because c could be negative, $a = b$ is equivalent to $a - c = b - c$.

> **Additive Property of Equality**
> For any real numbers a, b, and c, the equation
>
> $$a = b$$
>
> is equivalent to the equation
>
> $$a - c = b - c$$

Let's return to that example again.

EXAMPLE: Subtract to solve
Find the solutions to $x + 7 = 9$.

SOLUTION
We already know that the solution is $x = 2$, but let's see how we can figure this out in a systematic way.
   Rewrite the equation:

$$x + 7 = 9$$

We want to get rid of the 7 to isolate the variable on the left side of the equation. To do that, we add the *additive inverse* of 7, which is –7, to both sides of the equation:

$$x + 7 - 7 = 9 - 7$$

We do this so that –7 cancels out the 7 and leaves the variable isolated:

$$x + 0 = 9 - 7$$

or

$$x = 2$$

The original equation is equivalent to the final equation, which means that $x + 7 = 9$ and $x = 2$ have the same solutions. The only solution to $x = 2$ is 2, so 2 is the only solution to $x + 7 = 9$.

> ## KEY POINTS
>
> *Additive Inverse* A pair of numbers that differ only by a sign, like 2 and –2, are called **additive inverses**, because the result of *adding* them together is zero. Because the additive inverse of any number has the opposite sign as the number, it's sometimes also called the *opposite*.
>
> The additive inverse of any number *a* is –*a*: the minus sign flips the sign of *a*.
>
> The additive inverse of 0 is 0 itself.

Let's go through a few more examples.

EXAMPLE: Add to solve

Find all solutions to $1 = x - 4$.

SOLUTION

Write down the equation:

$$1 = x - 4$$

Isolate the variable by adding the additive inverse of the constant that's on the same side of the equation. In this case, the constant near x is –4; its additive inverse is 4:

$$1 + 4 = x - 4 + 4$$

Now simplify both sides to get the solution:

$$5 = x$$

The equation $1 = x - 4$ is equivalent to $x - 4 = 1$. So if you feel more comfortable with keeping the variable on the left side of the equation, flip the equation.

So far, we've used the additive property of equality only with constants. But it's equally fine to use it with variables. Take a look at this example:

EXAMPLE: Add variable terms to solve
Solve the equation $4x = 6 + 3x$.

SOLUTION
Rewrite the equation:

$$4x = 6 + 3x$$

Ultimately, we want to get the x terms on one side of the equation and the constant term on the other. Since there is only one constant term, on the right-hand side, we want to clear the variable term from that side by adding the additive inverse of $3x$ to both sides of the equation:

$$4x + (-3x) = 6 + 3x + (-3x)$$

Now simplify both sides:

$$4x - 3x = 6 + 0$$
$$x = 6$$

So $x = 6$ is the solution to $4x = 6 + 3x$.

To simplify things, you can write $-3x$ right away instead of $+(-3x)$. This cuts out some clutter; the solution then looks like this:

Rewrite the equation:

$$4x = 6 + 3x$$

Subtract $3x$ from both sides:

$$4x - 3x = 6 + 3x - 3x$$

Simplify:

$$x = 6$$

OPTIONS

Bypassing the Additive Property of Equality In practice, the additive property of equality is tantamount to the following principle: you can move a term from one side of the equation to the other if you flip its sign.

Take a look at what this means for the three examples that we've looked at.

EXAMPLE:

In $x + 7 = 9$, move the 7 from the left-hand side to the right-hand side by flipping its sign: $x = 9 - 7$. Then simplify to get $x = 2$.

EXAMPLE:

In $1 = x - 4$, move the -4 from the right-hand side to the left-hand side by flipping its sign: $1 + 4 = x$. Then simplify to get $5 = x$.

EXAMPLE:

In $4x = 6 + 3x$, move the $3x$ from the right-hand side to the left-hand side by flipping its sign: $4x - 3x = 6$. Then simplify to get $x = 6$.

SOLVING BY MULTIPLYING

The multiplicative property of equality works like the additive property of equality:

KEY POINTS

Multiplicative Property of Equality For any real numbers a, b, and nonzero real number c, the equation

$$a = b$$

is equivalent to the equation

$$ac = bc$$

In other words, it's legal to multiply both sides by any constant c because it won't fundamentally change the equation. And because c may well be less than 1, it's also fine to divide both sides of an equation by the same nonzero number.

For any real numbers a, b, and nonzero real number c, the equation

$$a = b$$

is equivalent to the equation

$$\frac{a}{c} = \frac{b}{c}$$

In practice we apply the multiplicative property of equality the same way that we apply the additive property—to isolate an x on one side of the equation and move all the numbers to the other side.

Take a look at how this works.

> ## KEY POINTS
>
> *Multiplicative Inverse* The **multiplicative inverse** of a number is the number you have to multiply by to get 1. So 2 and $\frac{1}{2}$ are multiplicative inverses because $2 \cdot \frac{1}{2} = 1$, and $-\frac{3}{4}$ and $-\frac{4}{3}$ are multiplicative inverses because $\left(-\frac{3}{4}\right)\left(-\frac{4}{3}\right) = 1$.
>
> Every number *except zero* has a multiplicative inverse. The multiplicative inverse of any nonzero number a is $\frac{1}{a}$.
>
> In the context of fractions the multiplicative inverse is also known as the **reciprocal**: it's the fraction flipped over, top to bottom.

EXAMPLE: Solve by multiplying
Solve the equation $3x = 24$.

SOLUTION
Rewrite the equation:

$$3x = 24$$

To get rid of the 3 and isolate the x on the left-hand side of the equation, multiply both sides by the *multiplicative inverse*, or *reciprocal*, of 3, which is $\frac{1}{3}$:

$$3x \cdot \frac{1}{3} = 24 \cdot \frac{1}{3}$$

Now simplify both sides:

$$\frac{3}{3}x = \frac{24}{3}$$

$$x = 8$$

Check: $3 \cdot 8 = 24$, so $x = 8$ is the solution.

In the example above, the 3 is called the **coefficient** of the x term. The coefficient is the real number that the variable is being multiplied by. In general, if there is a coefficient (not equal to 1) on the x term, you'll want to multiply both sides of the equation by the reciprocal of the coefficient to clear it away.

EXAMPLE: Solve by multiplying, fraction

Find the solution to $\dfrac{1}{2} = \dfrac{x}{10}$.

SOLUTION

Rewrite the equation:

$$\frac{1}{2} = \frac{x}{10}$$

To isolate x on the right-hand side, multiply both sides by 10 (that's the multiplicative inverse of $\dfrac{1}{10}$, the coefficient on the x term):

$$\frac{1}{2} \cdot 10 = \frac{x}{10} \cdot 10$$

Simplify both sides:

$$\frac{10}{2} = \frac{10}{10}x$$

$$5 = x$$

Check: $\dfrac{1}{2} = \dfrac{5}{10}$, so the solution is $x = 5$.

EXAMPLE: Solve by multiplying, negative

Find the solution of $-2x = 7$.

SOLUTION

Rewrite the equation:

$$-2x = 7$$

To clear out the -2 from the x side of the equation, multiply both sides by $-\frac{1}{2}$ (that's the reciprocal of -2):

$$-2x \cdot \left(-\frac{1}{2}\right) = 7 \cdot \left(-\frac{1}{2}\right)$$

Now simplify:

$$\frac{-2}{-2}x = -\frac{7}{2}$$

$$x = -\frac{7}{2}$$

Check: $-2\left(-\frac{7}{2}\right) = 7$, so $x = -\frac{7}{2}$ is the solution.

To eliminate some clutter, we can divide both sides by -2 instead of multiplying by $-\frac{1}{2}$. The solution then looks like this:

Rewrite the equation:

$$-2x = 7$$

Divide both sides by -2:

$$\frac{-2}{-2}x = \frac{7}{-2}$$

Simplify:

$$x = -\frac{7}{2}$$

EXAMPLE: Solve by multiplying, negative fraction

Solve the equation $-\frac{3}{8}y = -\frac{9}{2}$.

SOLUTION

Rewrite the equation:

$$-\frac{3}{8}y = -\frac{9}{2}$$

Multiply both sides by the reciprocal of the y coefficient:

$$-\frac{3}{8}\left(-\frac{8}{3}\right)y = -\frac{9}{2}\left(-\frac{8}{3}\right)$$

Simplify:

$$\left(-\frac{3}{8}\right)\left(-\frac{8}{3}\right)y = -\left(-\frac{9}{2} \cdot \frac{8}{3}\right)$$

$$y = \frac{\cancel{9}^{3} \cdot \cancel{8}^{4}}{\cancel{2} \cdot \cancel{3}} = 12$$

Check: $-\frac{3}{8}(12) = -\frac{36}{8} = -\frac{9}{2}$, so the solution is $y = 12$.

This final equation can be solved with either the additive or the multiplicative property:

EXAMPLE: Option one: multiply to solve

Find the solution to $-z = 0.5$.

SOLUTION

Rewrite the equation:

$$-z = 0.5$$

Multiply both sides by –1 (that's the reciprocal of –1, the coefficient on the z term):

$$-z(-1) = 0.5(-1)$$

This simplifies to

$$z = -0.5$$

EXAMPLE: Option two: add to solve
Solve $-z = 0.5$ another way.

SOLUTION
Move $-z$ to the right side of the equation by flipping its sign:

$$-z = 0.5$$
$$0 = 0.5 + z$$

Do it again: move 0.5 to the left-hand side of the equation by flipping its sign:

$$-0.5 = z$$

So the solution, again, is $z = -0.5$.

SOLVING BY ADDING AND MULTIPLYING

Taken together, the additive property of equality and the multiplicative property of equality can solve any equation in the form $Ax + B = Cx + D$, where A, B, C, and D are real-number constants, and x is the variable.

First, we look at a few examples where the variable is on one side only, as in $Ax + B = D$.

EXAMPLE: Add, then multiply to solve
Solve the equation $4x + 3 = 31$.

SOLUTION
Rewrite the equation:

$$4x + 3 = 31$$

To isolate x on the left-hand side, we need to get rid of the 3 and divide through by 4. First, subtract 3 and simplify:

$$4x + 3 - 3 = 31 - 3$$
$$4x = 28$$

Then, divide by 4 and simplify:

$$\frac{4x}{4} = \frac{28}{4}$$
$$x = 7$$

Check: $4(7) + 3 = 31$, so the solution is $x = 7$.

If the equation doesn't start out in $Ax + B = D$ form, you may need to simplify first.

EXAMPLE: Simplify first

Find the value(s) of y that makes $9 = 3(5 - 2y) + 2y$ true.

SOLUTION

Rewrite the equation:

$$9 = 3(5 - 2y) + 2y$$

Simplify the right side:

$$9 = 15 - 6y + 2y$$
$$9 = 15 - 4y$$

Isolate the y term on the right side by subtracting 15:

$$9 - 15 = 15 - 4y - 15$$
$$-6 = -4y$$

Divide through by –4 to isolate y:

$$\frac{-6}{-4} = \frac{-4y}{-4}$$
$$1.5 = y$$

Check by plugging in 1.5:

$$9 \stackrel{?}{=} 3\big(5 - 2(1.5)\big) + 2(1.5) = 3(5 - 3) + 3 = 9$$

So $y = 1.5$ is the solution.

VARIABLE ON BOTH SIDES

Finally, let's look at examples where the variable appears on both sides of the equation.

EXAMPLE: Variable on both sides
Solve the equation $5x + 7 = 8x - 5$.

SOLUTION
Rewrite the equation:

$$5x + 7 = 8x - 5$$

Use the additive property of equality to move all the x terms to one side. Since the x coefficient on the right-hand side (8) is greater than the coefficient on the left-hand side (5), move the x terms to the right side of the equation. (This is an arbitrary choice, but it will keep numbers a little more positive.) Subtract $5x$ from both sides to clear the left-hand side:

$$5x + 7 - 5x = 8x - 5 - 5x$$
$$7 = 3x - 5$$

Move the constant term to the left-hand side to isolate the xs on the right:

$$7 + 5 = 3x - 5 + 5$$
$$12 = 3x$$

Finally, divide through by 3 to get the solution:

$$\frac{12}{3} = \frac{3x}{3}$$
$$4 = x$$

Check: $5(4) + 7 = 27 \stackrel{?}{=} 8(4) - 5 = 27$, so $x = 4$ is the solution.

OPTIONS

Which Side Should the Variables Go On? When solving
$5x + 7 = 8x - 5$ we moved the variables to the right side and
the constants to the left side. You can also do it the other way:
move the variables to the left side and the constants to the
right side.

Starting with $5x + 7 = 8x - 5$, you have to subtract $8x$ from
both sides to move the variables to the left and then subtract
7 from both sides to move the constants to the right. You end
up with $-3x = -12$, which means that $x = 4$. That's the same
answer as we got before. You can do the problem either way.

The only difference: you end up having to mess around
with negative numbers, which some people find annoying. If
you prefer to deal with $3x = 12$ rather than $-3x = -12$, you
might want to try to move the variables to the side with the
greater coefficient. For example, in $5x + 7 = 8x - 5$, the left
side has $5x$ and the right side has $8x$. Move the variables to
the right side to deal with $3x$ rather than $-3x$ and avoid the
minus signs.

What About Flipping the Equation?

Some teachers recommend that you flip an equation like
$5x + 7 = 8x - 5$ and work with $8x - 5 = 5x + 7$ instead. Then
you get to have it both ways: you can move the variables to the
left (which some people prefer because we're more used to
seeing them there), *and* you avoid the extra minus signs.
That's fine too and will also give you the right answer.

If the equation is not already in the form $Ax + B = Cx + D$, simplify both sides until that form emerges.

EXAMPLE: Variable on both sides; simplify first
Solve $2(4x - 7) + 14 = -12(6 - x)$.

SOLUTION
Rewrite the equation:

$$2(4x - 7) + 14 = -12(6 - x)$$

Simplify both sides:

$$8x - 14 + 14 = -72 + 12x$$
$$8x = 12x - 72$$

Move the variables over to one side of the equation and the constants to the other. Since there's no constant on the left side, we'll put the variables there.

$$8x - 12x = 12x - 72 - 12x$$
$$-4x = -72$$

Divide through by –4:

$$\frac{-4x}{-4} = \frac{-72}{-4} = 18$$

Check: The left side is

$$2(4(18) - 7) + 14 = 2(72 - 7) + 14 = 130 + 14 = 144$$

and the right side is

$$-12(6 - 18) = -12(-12) = 144$$

So $x = 18$ is the solution.

TRICKS

Some optional multiplicative tricks can help make the computations a little easier. They may be applied at any time in the process of solving an equation. However, they do increase the number of steps it takes to find the solution, so only use them if you're comfortable with arithmetic.

If the equation involves fractions, you can clear the denominators by multiplying both sides by the least common multiple (LCM) of all the denominators.

Least Common Multiple

The **least common multiple (LCM)** of two integers is the smallest number that is divisible by both. For example, 80 is a common multiple of 4 and 10, but their least common multiple is 20.

To find the LCM of two or more integers, first find their prime factorization. Each prime that appears in one of the integers must appear in the LCM. If a prime appears more than once in any integer, it must appear in the LCM the greatest number of times that it appears anywhere.

For example, to find the LCM of 90 and 24, factor both:

$$90 = 2 \cdot 32 \cdot 5 \quad \text{and} \quad 24 = 2^3 \cdot 3$$

The primes that make an appearance are 2, 3, and 5; these primes will all appear in their LCM. Since 2 appears with exponent 3 in 24, 2^3 must appear in the LCM. Since 3 appears with exponent 2 in 90, 3^2 must appear in the LCM. So the LCM is $2^3 \cdot 3^2 \cdot 5 = 360$.

EXAMPLE: Trick one: clear denominators

Solve $\dfrac{3}{10} + \dfrac{x}{15} = \dfrac{5}{6}$.

SOLUTION

Rewrite the equation:

$$\frac{3}{10} + \frac{x}{15} = \frac{5}{6}$$

At this point, the fractions are overwhelming, so let's clear the denominators. The denominators are $10 = 2 \cdot 5$, $15 = 3 \cdot 5$, and $6 = 2 \cdot 3$, so their LCM is $2 \cdot 3 \cdot 5 = 30$. Multiply both sides by 30 and simplify:

$$30\left(\frac{3}{10} + \frac{x}{15}\right) = 30 \cdot \frac{5}{6}$$

$$\frac{90}{10} + \frac{30x}{15} = \frac{150}{6}$$

$$9 + 2x = 25$$

Now, isolate the x term by subtracting 9 from both sides:

$$9 + 2x - 9 = 25 - 9$$

$$2x = 16$$

Finally, isolate x by dividing both sides by 2:

$$\frac{2x}{2} = \frac{16}{2}$$

$$x = 8$$

Check: $\dfrac{3}{10} + \dfrac{8}{15} = \dfrac{3 \cdot 3}{10 \cdot 2} + \dfrac{8 \cdot 2}{15 \cdot 2} = \dfrac{25}{30} = \dfrac{5}{6}$, so $x = 8$ is the solution.

If the equation involves decimals, you can multiply both sides by a high enough power of 10 to get rid of the decimal points.

EXAMPLE: Trick two: clear decimal points

Solve $0.2 - 0.08z = 0.304$.

SOLUTION

Rewrite the equation:

$$0.2 - 0.08z = 0.304$$

Every number involved has at most three digits after the decimal point, so let's move the decimal point three places to the right—that is, let's multiply both sides by 1000:

$$0.2 \cdot 1000 - 0.08z \cdot 1000 = 0.304 \cdot 1000$$
$$200 - 80z = 304$$

Isolate the z term on one side by subtracting 200 from both sides.

$$200 - 80z - 200 = 304 - 200$$
$$-80z = 104$$

Isolate z on one side by dividing through by -80:

$$\frac{-80z}{-80} = \frac{104}{-80}$$

$$z = -\frac{13}{10} = -1.3$$

Check: $0.2 - 0.08 \cdot (-1.3) = 0.2 + 0.104 = 0.304$, so $z = -1.3$ is the solution.

If every term in the equation is divisible by a whole number in an obvious way, you can divide through by that common factor to simplify computation.

EXAMPLE: Trick three: Factor out common factors
Solve $4000x - 18{,}000 = -2000$.

SOLUTION
Rewrite the equation:

$$4000x - 18{,}000 = -2000$$

Every term in the equation is divisible by 2000, so let's factor it out of both sides.

$$\frac{4000x}{2000} - \frac{18{,}000}{2000} = -\frac{2000}{2000}$$
$$2x - 9 = -1$$

Isolate the x term on one side by adding 9 to both sides:

$$2x - 9 + 9 = -1 + 9$$
$$2x = 8$$

Isolate x on one side by dividing both sides by 2:

$$\frac{2x}{2} = \frac{8}{2}$$
$$x = 4$$

Check: $4000 \cdot 4 - 18{,}000 = 16{,}000 - 18{,}000 = -2000$, so $x = 4$ is the solution.

COUNTING SOLUTIONS

Most of the equations that we've been working with have had exactly one solution, and that's what will happen most of the time. That's what happens when you reduce the original equation to an equation of the type $x = a$. But there are special cases when that's not possible. Take a look:

EXAMPLE: Many solutions
Solve $3(x - 3) + 2x = 5x - 9$.

SOLUTION
Rewrite the equation:

$$3(x - 3) + 2x = 5x - 9$$

Simplify both sides:

$$3x - 9 + 2x = 5x - 9$$
$$5x - 9 = 5x - 9$$

The statement $5x - 9 = 5x - 9$ is true no matter the value of x. So *all real numbers* are solutions to the original equation.

Just to complete the picture, let's see what happens if we continue working on $5x - 9 = 5x - 9$:

$$5x - 9 = 5x - 9$$

Move the variable terms to one side of the equation:

$$5x - 9 - 5x = 5x - 9 - 5x$$
$$-9 = -9$$

Again, $-9 = -9$ is a true statement for all values of x. If we want to, we can continue:

$$-9 + 9 = -9 + 9$$
$$0 = 0$$

At this point, there's really nothing else to do: $0 = 0$ is true for all values of x.

EXAMPLE: No solutions

Find all the solutions to $13 - 2y = 5(y + 3) - 7y$.

SOLUTION

Rewrite the equation:

$$13 - 2y = 5(y + 3) - 7y$$

Simplify both sides:

$$13 - 2y = 5y + 15 - 7y$$
$$13 - 2y = -2y + 15$$

Move the variables to one side:

$$13 - 2y + 2y = -2y + 15 + 2y$$
$$13 = 15$$

The equation $\cancel{13 = 15}$ is never true, regardless of the value of y, so we can say it has no solutions. Since the original equation is equivalent to $\cancel{13 = 15}$, the original equation has no solutions either.

KEY POINTS

How Many Solutions Does a Linear Equation Have? There are only three scenarios for linear equations:

1. **One solution:** The equation reduces to a unique solution. This is the most common scenario.

2. **No solutions:** The equation reduces to a false statement and so has no solutions. This is a special case.

3. **An infinite number of solutions:** The equation reduces to a true statement, so all real numbers are solutions. This is an even more special case.

Word Problems

A **word problem** is popular type of math-class exercise in which a math problem is disguised in a few prose sentences. The solver's basic task is three-fold: translate the words into a math problem, solve the math problem, and then translate the answer back into words again. We will add some details to this process, but the three basic tasks are always the same.

Word problems can be very tricky. The best way to get good at solving word problems is to practice, so let's get started.

We'll use the same step-by-step process to work through every problem.

1. **Restate what you know:** Take the word problem and trim the fat: restate the key pieces of information in a succinct way, getting rid of all extraneous details. Occasionally you'll need to include commonly known facts, for example, that there are seven days in a week or fifty-two cards in a deck.

2. **Choose a variable:** The best decision is often, but not always, to let the variable represent what the word problem is asking for.

3. **Translate what you know into an equation:** Use the variable you've chosen to translate what you know from step 1 into an equation in terms of that variable.

4. **Solve the equation:** Do the math—solve for the variable.

5. **Translate the answer back and check that it makes sense:** Interpret your math answer as an answer to the word problem. Always check that the answer makes sense. If it doesn't, go back and look over your work.

The first word problem has a typical setup: a story about two people who have different amounts of a particular item. You need to find out how many of this item each person has.

EXAMPLE: Word problem: sum and difference

Paul and Lynn are avid plant collectors. Together they have 729 distinct species of plants, but Lynn has 51 more distinct species than Paul. How many distinct species of plants does Paul have?

SOLUTION

1. **Restate what you know** The problem tells us a lot of things: Paul and Lynn are the main characters, they are avid plant collectors, they care not only about plants but also about distinct species of plants, etc. Most of it, including what *distinct species* happens to mean, doesn't matter for the problem. What matters are two facts:

   • Paul's plants + Lynn's plants = 729

   • Lynn's plants = 51 more than Paul's plants

2. **Choose a variable** The problem asks how many distinct species of plants Paul has, so we'll let

$$x = \text{number of Paul's plant species}$$

3. **Translate into an equation** We pulled two statements out of the word problem earlier. We'll use one of them to express Lynn's plants in terms of Paul's plants. The other will become the main equation.

   First, we'll translate the second statement to express Lynn's plants in terms of Paul's:

$$
\begin{array}{ccccc}
\text{Lynn's plants} & = & 51 \text{ more than} & \text{Paul's plants} \\
\downarrow & \downarrow\;\downarrow & \downarrow & \downarrow \\
\text{Lynn's plants} & = & 51 \quad + & x
\end{array}
$$

Next, we'll translate the second statement:

$$
\begin{array}{ccccc}
\text{Paul's plants} & + & \text{Lynn's plants} & = & 729 \\
\downarrow & \downarrow & \downarrow & \downarrow\;\downarrow \\
x & + & (51+x) & = & 729
\end{array}
$$

4. **Solve the equation** This is a linear equation, and we already know how to solve those:

$$x + (51 + x) = 729 \qquad \text{Rewrite equation}$$
$$2x + 51 = 729 \qquad \text{Simplify}$$
$$2x + 51 - 51 = 729 - 51 \qquad \text{Subtract 51 from both sides}$$
$$2x = 678 \qquad \text{Simplify}$$
$$\frac{2}{2}x = \frac{678}{2} \qquad \text{Divide both sides by 2}$$
$$x = 339 \qquad \text{Simplify}$$

5. **Translate back and check** We've found that x, the number of Paul's plant species, is 339.

Check if the answer makes sense: if Paul has 339 plant species, then Lynn has $339 + 51 = 390$ plant species, and together they have $339 + 390 = 729$ plant species. So everything checks out.

Some word problems don't come with a story. They're just problems about math concepts told in words. Here's a classic type about consecutive integers.

EXAMPLE: Word problem: consecutive integers

The sum of three consecutive integers is 87. What are the three integers?

SOLUTION

1. **Restate what you know** *Consecutive* means "coming one right after the other." So the problem says that:

   some integer + the next integer + the next one after that = 87

2. **Choose a variable** We're asked to find all three integers, so the variable could represent any of them. Let's let

   $$n = \text{the first integer}$$

3. **Translate into an equation** There's only one statement here, so we translate it directly:

   some integer + the next integer + the next one after that = 87
   $$\downarrow \qquad \downarrow \qquad \downarrow \qquad \downarrow \qquad\qquad \downarrow \qquad\qquad \downarrow\downarrow$$
   $$n \quad + \quad (n+1) \quad + \quad (n+1+1) \qquad = 87$$

4. **Solve the equation**

$n + (n+1) + (n+1+1) = 87$	Rewrite equation
$3n + 3 = 87$	Simplify
$3n + 3 - 3 = 87 - 3$	Subtract 3 from both sides
$3n = 84$	Simplify
$\dfrac{3}{3}n = \dfrac{84}{3}$	Divide both sides by 3
$n = 28$	Simplify

5. **Translate back and check** If n, the smallest integer, is 28, then the three integers are 28, 29, and 30.

   Check: the sum of 28, 29, and 30 is indeed 87.

Age problems are yet another word-problem staple. Be careful with matching up what age each character was/is/will be.

EXAMPLE: Word problem: ages

Anne-Leigh's grandmother is exactly three times as old as Anne-Leigh. Nine years ago, Anne-Leigh's age was five times less than her grandmother's age next year. How old is Anne-Leigh's grandmother now?

SOLUTION

1. **Restate what you know** We know two facts:

    • grandmother's age = 3 times Anne-Leigh's age

    • and Anne-Leigh's age nine years ago = (grandmother's age next year) divided by 5

2. **Choose a variable** We're asked to find Anne-Leigh's grandmother's age, so that's a candidate for a variable. It's a little easier to work with Anne-Leigh's age, though, so we'll let

    a = Anne-Leigh's age now

3. **Translate into an equation** We'll use one of the two statements to get a relationship between Anne-Leigh's age and her grandmother's age and the other to get an equation to solve.

    The first statement gives us an easy way to express the grandmother's age in terms of Anne-Leigh's age.

    Translate:

    grandmother's age = 3 times Anne-Leigh's age
    ↓   ↓↓ ↓   ↓
    grandmother's age = 3 × a

    Next, we'll translate the second statement:

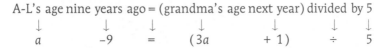

A-L's age nine years ago = (grandma's age next year) divided by 5
↓   ↓   ↓   ↓   ↓   ↓   ↓
a −9   =   ($3a$ + 1)   ÷   5

4. **Solve the equation**

$a - 9 = (3a + 1) \div 5$	Rewrite the equation
$(a - 9) \cdot 5 = (3a + 1) \div 5 \cdot 5$	Clear denominators
$5a - 45 = 3a + 1$	Simplify
$5a - 45 - 3a = 3a + 1 - 3a$	Move a terms to left side
$2a - 45 = 1$	Simplify
$2a - 45 + 45 = 1 + 45$	Move constants to right side
$2a = 46$	Simplify
$\dfrac{2}{2}a = \dfrac{46}{2}$	Isolate a on the left
$a = 23$	Simplify

5. **Translate back and check** We've found that a, or Anne-Leigh's age, is 23. This means that her grandmother is $3a$, or $3(23) = 69$, years old.

Check: Nine years ago, Anne-Leigh was 14, which is five times less than 70, the age her grandmother will be next year.

Summary

Expressions vs. equations

- An *expression* is a mathematical combination of numbers and variables. An *equation* says that two expressions are equal. Equations include equal signs; expressions don't. If you plug in values for the variables, an expression reduces to a number; an equation doesn't.

Order of operations

- First do parentheses, next exponential expressions, then multiplication and division, and last addition and subtraction.

Simplifying an expression

- A simplified expression involves no parentheses and has all its fractional expressions in lowest terms.

Solving equations

To solve a linear equation:

- Simplify both sides.
- Add the same quantities to both sides to isolate the variable terms on one side of the equation, with the constants on the other.
- Divide both sides by the coefficient of the variable to isolate the variable on one side of the equation.

Now you have the solution.

Counting solutions

- A linear equation is most likely to have exactly one solution. The next likeliest case is no solutions. Finally, some linear equations have all real numbers as solutions.

Word problems

- To solve a word problem, you *must* translate words into math, solve the math, and translate the math back into words.

  We *recommend* the following process: read the problem and restate the key information, choose your variable, translate the word problem into an equation, solve it, and translate back, checking that the answer is reasonable.

Sample Test Questions

Answers to these questions begin on page 385.

1. Evaluate each arithmetic expression.

    A. $(18 \cdot (12 \div 3) - (7 - 1)) \div (12 - (6 + 3))$

    B. $(10 - (9 - 5))^2 + \dfrac{-4^2}{7 - 5 \cdot 3}$

2. Evaluate each algebraic expression using the variable values provided.

    A. $17 - x(8 - x)$ for $x = -3$

    B. $(x + y + z)(x - y - z)$ for $x = 3$, $y = -4$, and $z = 5$

3. Simplify each of these expressions.

    A. $4a - 25 - 7a + 11$

    B. $3(5x + 1 - 2y)$

    C. $-4(m - 5) - 3(7 - 2m)$

    D. $\dfrac{1}{3}(p + q) - \dfrac{2}{3}(p - q)$

4. Ethel and Thelma are carrying a bag of apples home to their mother. On the way, Thelma eats half of all the apples. Ethel finds seven more, but Thelma eats a third of the new amount. Express the number of apples in their bag at the end of the journey in terms of a, where a is the number of apples in the bag at the beginning.

5.

    A. Is $x = 60$ a solution for $\dfrac{12}{5}x - 22 = 2x$?

    B. Is $x = 6$, $y = -5$ a solution for $6x + 7y = 1$?

6. Solve using the additive property of equality.

    A. $x - 17 = 12$

    B. $23 - (9 - x) = 4$

    C. $6x - 2 = 5x + 3$

7. Solve using the multiplicative property of equality.

    A. $-8y = -144$

    B. $\frac{2}{5}y = 38$

    C. $3(-5x) = 40$

8. Find all solutions to each equation.

    A. $3x - 11 = 46$

    B. $-4(3x - 5) = 50$

    C. $\frac{2}{3}x + \frac{1}{6}(x - 5) = \frac{3}{4}$

9. Solve each equation.

    A. $2(x - 5) + 2x = 4(x + 3) - 22$

    B. $2x + 7 = 5x + 28$

    C. $1.30x + 3.09 = 2.52x - 4.23$

    D. $-3x + 17 = 5(x - 3) - 2(4x + 1)$

    E. $4(7x + 9) = 8(2x + 1)$

10. David and Carl are freshman-year college roommates, each struggling with laundry for the first time. Together, they've shrunk a total of 28 articles of clothing, all T-shirts and underwear. If David has shrunk 3 times as many articles of clothing as Carl, how many pairs of underwear and T-shirts has David shrunk in the laundry during his freshman year so far?

11. Gerald has an important paper to write, so to avoid all distractions, he decides to spend the day working at his favorite coffee shop. Round-trip bus fare to the coffee shop costs $1.75, and Gerald figures that he'll have to spend at least $1.25 an hour on coffee and treats while he's there. If Gerald has $13, how many hours will he be able to work at the coffee shop today?

12. Ethel and Thelma are carrying a bag of apples home to their mother. On the way, Thelma eats half of all the apples. Ethel finds seven more, but Thelma eats a third of the new amount. When they get home, they find that they have two more apples than they started with. How many apples were in their bag at the beginning?

Graphing Linear Equations

2

Overview

One way of understanding equations is to draw pictures of their solutions. These pictures, or *graphs*, give us a visual understanding of the behavior of the equation. We humans like visualizing— it's much easier to remember a shape on a page than a string of constants and coefficients. Graphing is a way of capitalizing on the intuitions of geometry to solve algebraic problems.

Graphing One-Variable Equations

Solutions to one-variable equations are just numbers, so you can plot them on the real number line.

For example, the only solution to $2(x + 7) = 5$ is $x = -4.5$, so it's easy to plot all the solutions to the equation:

$2(x + 7) = 5$

This picture of all the solutions to the equation $2(x + 7) = 5$ plotted on the number line is called the **graph** of this equation. The graph of $2(x + 7) = 5$, like the graph of most one-variable linear equations, is a single point.

We can **graph** more-complicated one-variable equations as well. For example, the equation $x^2 = 9$ has exactly two solutions, $x = 3$ and $x = -3$. So $x^2 = 9$ graphed on the number line is just two points:

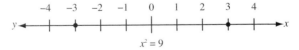

$x^2 = 9$

Two-Variable Equations

A *two-variable equation* is an equation that involves two variables, like $2x - y = 3$ or $y^2 = x + 1$.

A *solution* to an equation in two variables is a pair of values, one for each variable, that makes the equation true. For example, one solution to $y^2 = x + 1$ is $x = 15$, $y = 4$, because $(4)^2 = (15) + 1$.

ORDERED PAIRS

For convenience, a solution like $x = 15$, $y = 4$ to $y^2 = x + 1$ can also be written as $(15, 4)$, an **ordered pair**. By convention, the variable values in an ordered pair come alphabetically: the solution $(-2, 1)$ for the equation $a + 5b = 3$ means that $a = -2$ and $b = 1$, not the other way around.

In particular, for an equation in x and y, the x-value comes first and the y-value comes second. In fact, x and y are so popular as variables that the two values, or **coordinates**, in an ordered pair are often called the **x-coordinate** and the **y-coordinate**.

If there's any confusion, you can clarify: $(a, k) = (0, 7)$ means that $a = 0$ and $k = 7$.

KEY POINTS

Ordered Pair In an *ordered* pair, the *order* of the coordinates matters. For example, $(2, -3)$ is not the same ordered pair as $(-3, 2)$. Two ordered pairs are equal if and only if their first coordinates are equal and their second coordinates are equal, as real numbers.

As a point of comparison, *sets* are not ordered. The two-element set $\{2, -3\}$ is identical to $\{-3, 2\}$.

The Coordinate Plane

The number line is **one-dimensional**: to identify a point on it, you need to give just one number. One dimension, one variable.

To graph solutions to a two-variable equation in a useful way, we'll need some kind of **two-dimensional** system—for example, a plane.

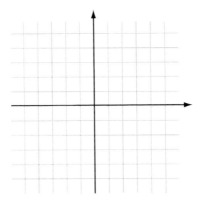

The **coordinate plane**, also known as the **Cartesian plane** and the **xy-plane**, is a system for identifying points on a plane. It consists of two perpendicular number lines, called **axes**, intersecting at the **origin**, the point where both number lines are 0. By convention we usually draw a horizontal **x-axis** whose numbers increase from right to left and a vertical **y-axis** whose numbers increase from bottom to top.

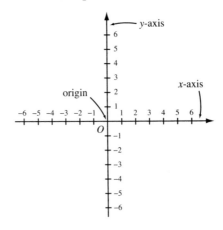

A point on the plane is identified by an ordered pair whose first (or x-) coordinate is the perpendicular distance from the y-axis and whose second (or y-) coordinate is the perpendicular distance from the x-axis. Here is a coordinate plane with the points $P = (3, 1)$, $Q = (2, 7)$, $R = (-3, -4)$, and $S = (-5, 2)$.

The origin, commonly labeled O, is the point $(0, 0)$.

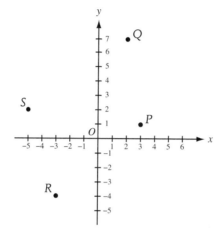

EXAMPLE: Name points

Identify each point in the plane by its ordered pair.

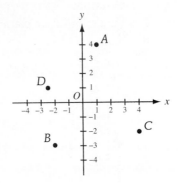

SOLUTION

Point *A* is 1 unit to the right of the *y*-axis and 4 units above the *x*-axis. Its coordinates are $(1, 4)$.

Point *B* is 2 units to the left of the *y*-axis and 3 units below the *x*-axis. Its coordinates are $(-2, -3)$.

Point *C* is 4 units to the right of the *y*-axis and 2 units below the *x*-axis. Its coordinates are $(4, -2)$.

Point *D* is $2\frac{1}{2}$ units to the left of the *y*-axis and 1 unit above the *x*-axis. Its coordinates are $(-2.5, 1)$.

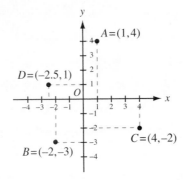

EXAMPLE: Plot ordered pairs

Plot each of these points on the coordinate plane.

a. $A = (7, 3)$

b. $B = \left(\dfrac{1}{2}, 1\right)$

c. $C = (-2, 0)$

d. $D = (0, -4)$

SOLUTION

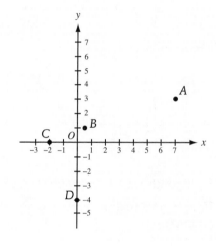

QUADRANTS AND HALF-PLANES

Quadrants

The two axes divide the plane into four **quadrants**, which are commonly numbered as in the figure below. The first quadrant is in the upper right; the others follow counterclockwise.

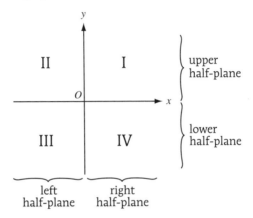

- **Quadrant I:** all points with two positive coordinates
- **Quadrant II:** all points with a negative x-coordinate and a positive y-coordinate
- **Quadrant III:** all points with two negative coordinates
- **Quadrant IV:** all points with a positive x-coordinate and a negative y-coordinate

By convention, points on an axis are not included in any quadrant.

EXAMPLE: Quadrants
Identify which quadrant each point belongs to.

a. $A = (-17, 4)$

b. $B = (25, -18)$

c. $C = (-31, 92)$

d. $D = (42, 72)$

e. $E = (-12, -86)$

f. $F = (6, 0)$

SOLUTION

a. Point A has a negative x-coordinate and a positive y-coordinate, so it's in quadrant II.

b. Point B has a positive x-coordinate and a negative y-coordinate, so it's in quadrant IV.

c. Point C has a negative x-coordinate and a positive y-coordinate, so it's in quadrant II.

d. Point D has two positive coordinates, so it's in quadrant I.

e. Point E has two negative coordinates, so it's in quadrant III.

f. Point F has a zero y-coordinate, so it's on the x-axis and doesn't belong to any quadrant.

Half-Planes

The axes also divide the plane into half-planes, which are sometimes called by the following self-explanatory names:

- **Upper half-plane:** all points above the x-axis; that is, quadrants I and II along with the positive y-axis
- **Lower half-plane:** all points below the x-axis; that is, quadrants III and IV and the negative y-axis
- **Right half-plane:** all points to the right of the y-axis; that is, quadrants I and IV and the positive x-axis
- **Left half-plane:** all points to the left of the y-axis; that is, quadrants II and III and the negative x-axis

DISTANCE IN THE PLANE

We now have a system for identifying points in the plane. How can we tell how far apart they are?

The distance between two points on a one-dimensional number line is the absolute value of their difference: the distance between a and b is $|b - a|$. (We need to take the absolute value because we don't know which of a and b is greater, and we don't want to get a negative distance.)

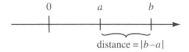

In the plane, two points that have the same y-coordinate lie on a horizontal line, which is similar to a number line. So the distance between them is just the absolute value of the difference in their x-coordinates: the distance between (a_1, b) and (a_2, b) is $|a_2 - a_1|$.

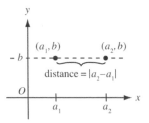

Similarly, we can turn the whole picture around 90 degrees. Two points that have the same x-coordinate lie on a vertical line, so the distance between them is the absolute value of the difference in their y-coordinates: the distance between (a, b_1) and (a, b_2) is $|b_2 - b_1|$.

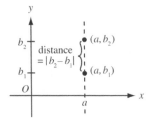

But if two points have different x- and y-coordinates, the distance between them is trickier to compute.

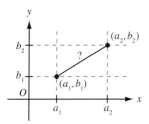

We need some way of determining distances along diagonals. For this, we need a formula from geometry.

Pythagorean Theorem

To review, a **right triangle** is a triangle with a right (or 90°) angle and, thus, two perpendicular sides. These two sides are called the **legs**; the third and longest side is the **hypotenuse**.

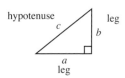

The Pythagorean Theorem is a formula that allows us to find the hypotenuse of a right triangle if we know the lengths of the legs.

> *Pythagorean Theorem*
>
> The square of the length of the hypotenuse of a right triangle is equal to the sum of the squares of the two legs.
>
> If a and b are the lengths of the two legs of a right triangle and c is the length of the hypotenuse, then
>
> $$a^2 + b^2 = c^2$$
>
> Equivalently, $c = \sqrt{a^2 + b^2}$.

EXAMPLE: Pythagorean Theorem

Find the length of the hypotenuse given the length of the legs in each right triangle.

a. $a = 3, b = 4$

b. $a = 6, b = 8$

c. $a = 12, b = 5$

d. $a = 8, b = 15$

e. $a = 2, b = 3$

SOLUTION

a. $c = \sqrt{a^2 + b^2} = \sqrt{3^2 + 4^2} = \sqrt{25} = 5$

b. $c = \sqrt{6^2 + 8^2} = \sqrt{100} = 10$

c. $c = \sqrt{12^2 + 5^2} = \sqrt{169} = 13$

d. $c = \sqrt{8^2 + 15^2} = \sqrt{289} = 17$

e. $c = \sqrt{2^2 + 3^2} = \sqrt{13}$

Distance from the Origin

We can use the Pythagorean theorem to determine the distance from a point to the origin.

A point (a, b) forms a right triangle with the x-axis and the origin. Here's what it looks like if (a, b) is in quadrant I, where a and b are both positive:

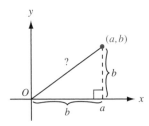

The two legs of this right triangle have lengths exactly a and b, and the hypotenuse is the distance of the point from the origin. So (a, b) is exactly $\sqrt{a^2 + b^2}$ units away from the origin.

You can make a similar right-triangle diagram in any other quadrant, where a or b or both are negative. But since you end up squaring a and b, it turns out that the negative signs don't make a difference. The same formula works for any point (a, b) in the coordinate plane.

> *Distance from the Origin*
>
> The distance from point (a, b) to the origin is given by
>
> $$d = \sqrt{a^2 + b^2}$$

EXAMPLE: Distance to origin

In each case, find the distance from P to the origin.

a. $P = (1, 1)$

b. $P = (-2, -5)$

c. $P = (-3, 4)$

d. $P = (6, 0)$

SOLUTION

Use the formula to find the distance in each case.

a. Distance $= \sqrt{1^2 + 1^2} = \sqrt{2}$

b. Distance $= \sqrt{(-2)^2 + (-5)^2} = \sqrt{29}$

c. Distance $= \sqrt{(-3)^2 + 4^2} = \sqrt{25} = 5$

d. Distance $= \sqrt{6^2 + 0^2} = \sqrt{36} = 6$. In this case, we didn't have to use the formula: the distance from the origin is along the x-axis and therefore easy to determine.

Distance Formula

Using similar methods, we get the general distance formula between two points in the plane.

> **Distance Formula**
>
> The distance between points (a_1, b_1) and (a_2, b_2) is given by
>
> $$d = \sqrt{(a_2 - a_1)^2 + (b_2 - b_1)^2}$$

Note that if $b_1 = b_2$ (that is, if the two points have the same y-coordinate), we get $d = \sqrt{(a_2 - a_1)^2}$. It turns out that that's exactly equal to $|a_2 - a_1|$, which we got for the distance between two points with the same y-coordinate. Similarly, if $a_1 = a_2$ we get $d = |b_2 - b_1|$, which is what we got for the distance between two points with the same x-coordinate.

EXAMPLE: Distance formula
In each case, find the distance between P and Q.

a. $P = (1, 2)$ and $Q = (5, 5)$
b. $P = (-9, 4)$ and $Q = (3, -1)$

SOLUTION
Use the distance formula.

a. Distance $= \sqrt{(5-1)^2 + (5-2)^2} = \sqrt{4^2 + 3^2} = 5$
b. Distance $= \sqrt{(3-(-9))^2 + ((-1)-4)^2}$
$= \sqrt{12^2 + (-5)^2} = 13$

Graphing by Plotting Points

Let's return to the problem of plotting solutions to a two-variable equation, say, $2x - y = 3$. We already have a way of plotting each individual point, so let's find a few solutions and see what happens.

FINDING SOLUTIONS ALGEBRAICALLY

It's usually easy to find solutions when x or y is 0.

If $x = 0$, then $2x - y = 3$ becomes $2(0) - y = 3$, or $-y = 3$. So $y = -3$, and $(0, -3)$ is a solution.

If $y = 0$, then $2x - y = 3$ becomes $2x - 0 = 3$, or $2x = 3$. So $x = \dfrac{3}{2} = 1.5$, and $(1.5, 0)$ is a solution.

So far, so good. Let's find a few more. One way of generating more solutions is to pick a few values for x and use them to find corresponding values for y. We already know that $(0, -3)$ is a solution, so let's try $x = 1$, $x = 2$, $x = 3$, and $x = 4$. We can store our findings in a table:

x	y
0	-3
1	
1.5	0
2	
3	
4	

When $x = 1$, the equation becomes $2(1) - y = 3$, or $2 - y = 3$. So $2 - 3 = y$, or $y = -1$.

When $x = 2$, the equation becomes $2(2) - y = 3$, or $4 - y = 3$. So $4 - 3 = y$, or $y = 1$.

When $x = 3$, the equation becomes $2(3) - y = 3$, or $6 - y = 3$. So $6 - 3 = y$, or $y = 3$.

When $x = 4$, the equation becomes $2(4) - y = 3$, or $8 - y = 3$. So $8 - 3 = y$, or $y = 5$.

x	y
0	-3
1	-1
1.5	0
2	1
3	3
4	5

EXAMPLE: Find solutions algebraically I

Find five solutions to the equation $y = 3x$.

SOLUTION

The equation gives y as an expression in terms of x, so the easiest thing to do is pick values of x and compute corresponding values of y. We'll pick the x-values –2, –1, 0, 1, and 2, just to keep the values small. But any five values of x will do.

x	y
-2	-6
-1	-3
0	0
1	3
2	6

EXAMPLE: Find solutions algebraically II

Find five solutions to the equation $y = -\dfrac{1}{2}x + 3$.

SOLUTION

Again, the form of the equation makes it easy to find y once we have a value of x in mind, so that's what we'll do. Because the value of x will have to be multiplied by $\dfrac{1}{2}$, we'll also choose *even* numbers: let's go with –4, –2, 0, 2, and 4.

x	y
–4	5
–2	4
0	3
2	2
4	1

PLOTTING SOLUTIONS

Here's the coordinate plane with the six solutions to $2x - y = 3$ that we found in the previous section.

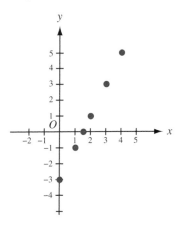

The solution points all lie on a straight line.

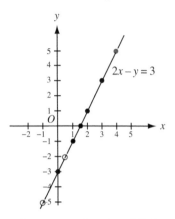

This line goes through many points in addition to the six that we already plotted. For example, it goes through (–1, –5) and through (0.5, –2). What happens if we plug in the coordinates of these points into the equation?

If we plug in $x = -1$ and $y = -5$, we get $2(-1) - (-5) = 3$, which is a true statement. So (–1, –5) is also a solution of the equation.

If we plug $x = 0.5$ and $y = -2$, we get $2(0.5) - (-2) = 3$, which is also a true statement. So (0.5, –2) is a solution to the equation as well.

In fact, it turns out that *every single point* on this line is a solution to $2x - y = 3$. Not only that, but every single solution to $2x - y = 3$ lies on this line. The set of all points in the plane that are solutions to a particular two-variable equation is called the *graph* of that equation. So this line that we've found is the *graph* of the equation $2x - y = 3$.

EXAMPLE: Find solutions from graph

Use the graph of $2x - y = 3$ to find a solution when

a. $x = 6$

b. $y = -4$

Verify that the ordered pairs you find are indeed solutions. Use the graph to find one more solution to this equation.

SOLUTION

Follow the line of the graph to find the points $(6, 9)$ and $\left(-\frac{1}{2}, -4\right)$.

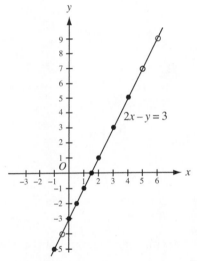

To check that these are solutions, plug them into the equation: $2(6) - 9 = 3$, so $(6, 9)$ is a solution. $2\left(-\frac{1}{2}\right) - (-4) = -1 + 4 = 3$, so $\left(-\frac{1}{2}, -4\right)$ is a solution.

Another point on the graph is $(5, 7)$. Since $2(5) - 7 = 3$, this point is also a solution to $2x - y = 3$.

EXAMPLE: Graph and find solutions I

Plot the five solutions that you found to the equation $y = 3x$ (see example on page 66). Do they lie on a straight line? If so, draw the line and use it to find two more solutions to $y = 3x$.

SOLUTION

We found these solutions: $(-2, -6)$, $(-1, -3)$, $(0, 0)$, $(1, 3)$, and $(2, 6)$. The five points found in the example do indeed lie on a straight line.

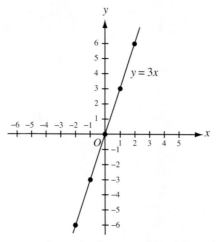

The line also passes through, for example, $(-3, -9)$ and $(3, 9)$. Both of these are also solutions: $3(-3) = -9$ and $3(3) = 9$.

It turns out that this line is a graph of the equation $y = 3x$.

LINEAR EQUATIONS

The **graph** of a two-variable equation is the set of points (x, y) in the coordinate plane that are solutions to the equation.

It turns out that the graph of every equation that can be written in the form $Ax + By = C$ is a straight line (caveat: A and B cannot both be zero). And vice versa—every straight line in the coordinate plane is the graph of some equation in the form $Ax + By = C$. That's why equations that can be reduced to the form $Ax + By = C$, with A and B not both zero, are called **linear equations**.

EXAMPLE: Graph and find solutions II

Plot the five solutions that you found to the equation
$y = -\frac{1}{2}x + 3$ (see example on page 67). Do they lie on a
straight line? If so, draw the line and use it to find two more
solutions to $y = -\frac{1}{2}x + 3$.

SOLUTION

We found these solutions: $(-4, 5)$, $(-2, 4)$, $(0, 3)$, $(2, 2)$, and
$(4, 1)$. These five points do indeed lie on a straight line:

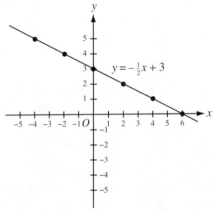

The line also passes through, for example, $(6, 0)$ and $\left(1, \frac{5}{2}\right)$.
Both of these are also solutions: $-\frac{1}{2}(6) + 3 = 0$ and
$-\frac{1}{2}(1) + 3 = \frac{5}{2}$.

It turns out that this line is the graph of the equation
$y = -\frac{1}{2}x + 3$.

An equation need not actually *be* in an $Ax + By = C$ form to be
called linear or have a straight-line graph; it merely needs to be
equivalent to an equation of that form. Equivalent equations have
the same solutions and so have the same graph.

EXAMPLE: Convert to standard form

Show that $y + 3x = 4(x + 7) - y$ is a linear equation by producing an equivalent equation in $Ax + By = C$ form.

SOLUTION

Use the same tried-and-true methods: eliminate all parentheses, combine all like terms, move all the variables to one side, and combine like terms again.

$$
\begin{aligned}
y + 3x &= 4(x + 7) - y \\
y + 3x &= 4x + 28 - y \\
y + 3x - 4x &= 4x + 28 - y - 4x \\
y - x &= 28 - y \\
y - x + y &= 28 - y + y \\
-x + 2y &= 28
\end{aligned}
$$

The equation $-x + 2y = 28$ is in $Ax + By = C$ form. (Remember that $-x$ is just $(-1)x$, so that $A = -1$, $B = 2$, and $C = 28$.) Therefore, $-x + 2y = 28$ is a linear equation. Since $y + 3x = 4(x + 7) - y$ is equivalent to it, $y + 3x = 4(x + 7) - y$ is a linear equation as well.

Sometimes it is more important to know *whether* an equation is expressible in $Ax + By = C$ form than it is to actually be able to produce this form. With a little experience, you should be able to eyeball many equations to determine whether they are linear. Linear equations involve only constant multiples of the variables; they do not have x^2 or xy or other products-of-variables terms.

EXAMPLE: Is equation linear?

Which of the following equations are linear?

a. $y = \frac{3}{4}x - 7$

b. $y = 2x(x + 1) + 3$

c. $x = 4$

d. $y = 8$

e. $xy = 32$

SOLUTION

a. This is a linear equation: $y = \frac{3}{4}x - 7$ is equivalent to $-\frac{3}{4}x + y = -7$, which is in $Ax + By = C$ form with $A = -\frac{3}{4}$, $B = 1$, and $C = -7$.

b. This is not a linear equation. If you multiply out the right side, you get $y = 2x \cdot x + 2x + 3$. The $2x \cdot x$ term is equal to $2x^2$, which makes the equation nonlinear.

c. This is a linear equation. It is in $Ax + By = C$ form with $A = 1$, $B = 0$, and $C = 4$.

d. This is a linear equation. It is in $Ax + By = C$ form with $A = 0$, $B = 1$, and $C = 8$.

e. This is not a linear equation. The term xy on the left-hand side is a product of variables, which can never appear in a linear equation. No matter how you manipulate the equation $xy = 32$, you won't be able to separate the x and the y by a plus or minus sign, as you'd need to in order to get it into the form $Ax + By = C$.

MORE LINEAR GRAPHS

When plotting the graph of a linear equation, you should always find at least three points—two to make the line and one more to check. If the three points don't lie on the same line, then you know you've made a mistake.

EXAMPLE: 3-point plot I

Plot the graph of $y = \dfrac{3}{4}x - 2$.

SOLUTION

First make a table of three solution points. Since the coefficient on x in the equation is $\dfrac{3}{4}$, we choose values of x that are multiples of 4—say −4, 0, and 4.

x	y
−4	−5
0	−2
4	1

Now plot:

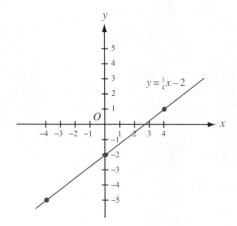

EXAMPLE: 3-point plot II

Plot the graph of $y = -\dfrac{5}{2}x + 1$.

SOLUTION

First make a table of three solution points. Since the coefficient of x in the equation is $-\dfrac{5}{2}$, we choose values of x that are multiples of 2: say, –2, 0, and 2.

x	y
–2	6
0	1
2	–4

Now plot:

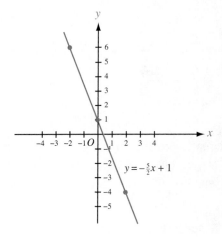

$y = -\frac{5}{2}x + 1$

EXAMPLE: Line y = b

Plot the graph of $y = 2$.

SOLUTION

In this equation, the value of x doesn't matter: every solution must have $y = 2$. If we choose 0, 1, and 2 for x-values, we'll get these solution points:

x	y
0	2
1	2
2	2

The graph is a horizontal line.

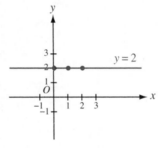

In fact, every equation in the form $y = b$ has a horizontal line at height b as its graph. If b is positive, the line is above the x-axis; if b is negative, the line is below the x-axis. The graph of $y = 0$ is the x-axis itself.

> **Graph of y = b**
> The graph of
> $$y = b$$
> is a horizontal line at height b.

EXAMPLE: Line x = a

Plot the graph of $x = -3$.

SOLUTION

In this equation, the value of *y* doesn't matter: every solution must have $x = -3$. If we choose 0, 1, and 2 for *y*-values, we'll get these solutions points:

x	y
-3	0
-3	1
-3	2

The graph is a vertical line.

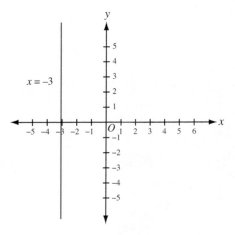

The same thing happens for any equation in the form $x = a$: its graph is a vertical line through $(a, 0)$. If *a* is positive, the line is to the right of the *y*-axis; if *a* is negative, the line is to the left of the *y*-axis. The graph of $x = 0$ is the *y*-axis itself.

> *Graph of x = a*
> The graph of
>
> $$x = a$$
>
> is a vertical line through $(a, 0)$.

Lines in the Coordinate Plane

First, let's review some terms from plane geometry.

GEOMETRY OF LINES

Two points in the plane determine a line. This means that given any two points, there is one and exactly one line that goes through them.

Three or more points lying on the same line are said to be **colinear**.

Two lines in the plane either intersect or they are **parallel**. Parallel lines never intersect.

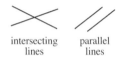

intersecting   parallel
lines   lines

Given a line and a point not on the line, there is exactly one line through the point parallel to the original line.

Two lines that intersect at right angles are called **perpendicular**.

perpendicular
lines

Given a line and a point on the line, there is exactly one line perpendicular to the line that passes through the point. Also, given a line and a point off the line, there is exactly one perpendicular that passes through that point.

SLOPE OF A LINE

The **slope** of a line in the coordinate plane is a measure of its steepness. If (x_1, y_1) and (x_2, y_2) are two points on the line, then the slope of the line is the ratio of the change in y to the change in x.

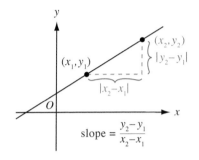

Slope of a line

$$\frac{\text{change in } y}{\text{change in } x} = \frac{y_2 - y_1}{x_2 - x_1}$$

Sometimes "change in y over change in x" is called "rise over run."

EXAMPLE: Slope from graph, positive

Compute the slope of the line.

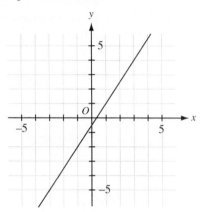

SOLUTION

This line goes through many points, including $(-1, -2)$, $(1, 1)$, and $(3, 4)$. Pick two of them, say, $(1, 1)$ and $(3, 4)$, and compute:

$$\frac{\text{change in } y}{\text{change in } x} = \frac{4 - 1}{3 - 1} = \frac{3}{2}$$

So the slope of this line is $\frac{3}{2}$.

Work through the examples in this section, keeping in mind that the slope of a line is the same no matter which two points you happen to pick to perform the calculation. The order of the two points also doesn't matter, so long as you're consistent in the numerator and the denominator. So if we choose another pair of points, say $(3, 4)$ and $(-1, -2)$, in the example above, we'll get the same slope:

$$\frac{\text{change in } y}{\text{change in } x} = \frac{-2 - 4}{-1 - 3} = \frac{-6}{-4} = \frac{3}{2}$$

EXAMPLE: Slope from graph, negative

Find the slope of the line.

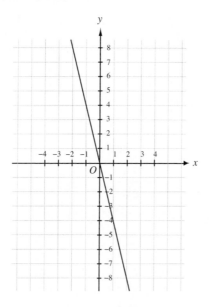

SOLUTION

Choose two points on the line—for example, $(-2, 8)$ and $(0, 0)$. The slope is

$$\frac{\text{change in } y}{\text{change in } x} = \frac{0 - 8}{0 - (-2)} = -4$$

Positive and Negative Slope

As you've already seen, slope may be a positive or a negative number. Lines with positive slope go up as they go right (equivalently, down and left); change in y and change in x have the same sign. Lines with negative slope go down as they go right (equivalently, up and left); change in y and change in x have different signs.

EXAMPLE: Slope, positive and negative

Compute the slope of these two lines.

a.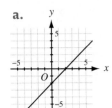

b.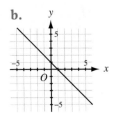

SOLUTION

a. The line goes through $(0, -2)$ and $(2, 0)$, among other points. Its slope is $\dfrac{0-(-2)}{2-0} = \dfrac{2}{2} = 1$.

b. The lines goes through $(0, 1)$ and $(1, 0)$, among other points. Its slope is $\dfrac{0-1}{1-0} = \dfrac{-1}{1} = -1$.

These two lines have the same steepness, but one has positive slope and the other negative slope.

EXAMPLE: Sign of slope from description

Determine whether the slope is positive or negative, without computing it.

a.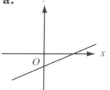

b.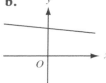

c.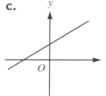

d.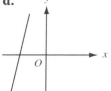

e.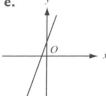

f.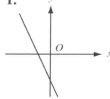

SOLUTION

a. Line moves up right; positive slope

b. Line moves down right; negative slope

c. Line moves up right; positive slope

d. Line moves up right; positive slope

e. Line moves up right; positive slope

f. Line moves down right; negative slope

Slope of Horizonal and Vertical Lines

Let's determine the slope of a horizontal line.

EXAMPLE: Horizontal line
Find the slope of this line.

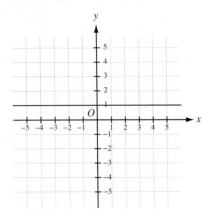

SOLUTION
The line goes through, for example, $(0, 1)$ and $(1, 1)$. So its slope is $\dfrac{1-1}{1-0} = \dfrac{0}{1} = 0$.

The slope of this line is zero.

The same thing will happen for any horizontal line: change in y is always zero, and zero divided by any nonzero number is zero.

> **Slope of a Horizontal Line**
> The slope of a horizontal line is 0.

Now, let's find the slope of a vertical line.

EXAMPLE: Vertical line
Find the slope of the line.

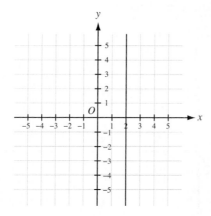

SOLUTION
The line goes through, for example, (2, 1) and (2, 2). So its slope is

$$\frac{2-1}{2-2} = \frac{1}{0} = \dots$$

You can't divide by zero, so you can't compute the slope of this line. Its slope is *undefined*.

The same thing will happen for any vertical line: change in x is always zero, and dividing by zero is not allowed.

> *Slope of a Vertical Line*
> The slope of a vertical line is undefined.

Large and Small Slope

Steep lines have slope whose absolute value is large, certainly greater than 1. Shallow lines have slope whose absolute value is very small, certainly less than 1.

EXAMPLE: Slope large and small
Find the slope of these lines.

a.

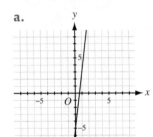

b.

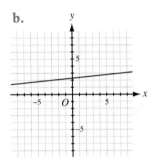

SOLUTION

a. The line goes through $(0, -5)$ and $(1, 3)$ so its slope is
$$\frac{3 - (-5)}{1 - 0} = 8.$$

b. The lines goes through $(7, 3)$ and $(-3, 2)$, so its slope is
$$\frac{2 - 3}{-3 - 7} = \frac{1}{10}.$$

Slopes of Parallel and Perpendicular Lines

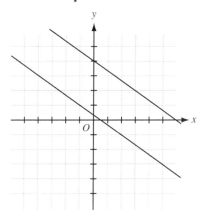

The figure above shows two parallel lines. What are their slopes?

The top line goes through (0, 4) and (3, 2); its slope is

$$\frac{2-4}{3-0} = -\frac{2}{3}$$

The bottom line goes through (–1, 1) and (2, –1); its slope is

$$\frac{-1-1}{2-(-1)} = -\frac{2}{3}$$

Their slopes are the same.

In fact, parallel lines always have the same slope—and vice versa: two lines that have the same slope are parallel. This makes sense: parallel lines point in the same direction, and slope is just a numerical way of keeping track of direction.

> *Parallel Lines*
> Parallel lines have the same slope.

The flip side of this observation is that knowing the slope of a line is not enough to identify it uniquely, since any number of lines parallel to the same line have the same slope. To identify a line, you also need to know how high or low it runs—for example, where it crosses the y-axis.

KEY POINTS

Perpendicular Lines Perpendicular lines have slopes that are negative reciprocals of each other: if a line has slope m, a line perpendicular to it will have slope $-\dfrac{1}{m}$.

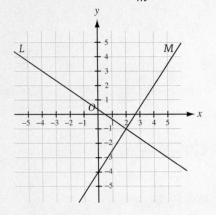

In the image above, line L goes through $(-1, 1)$ and $(2, -1)$; its slope is $\dfrac{-1-1}{2-(-1)} = -\dfrac{2}{3}$. Line M goes through $(2, -1)$ and $(4, 2)$; its slope is $\dfrac{2-(-1)}{4-2} = \dfrac{3}{2}$. Since $-\dfrac{2}{3} = -\dfrac{1}{3/2}$, the slopes are negative reciprocals and the lines are perpendicular.

Drawing Graphs of Equations

In the previous section, we looked at the *slope* of a line in the coordinate plane. Now we get to use everything that we've learned about slope to draw graphs of linear equations quickly.

THE SLOPE IN THE EQUATION

Let's start by graphing an equation: $y = \frac{1}{2}x - 1$. We choose x-values –2, 0, and 2 and find the corresponding y-values:

x	y
-2	-2
0	-1
2	0

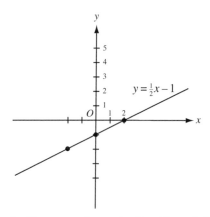

We know that the line goes through (0, –1) and (2, 0), so the slope is $\frac{0-(-1)}{2-0} = \frac{1}{2}$.

This is exactly the coefficient of x in the equation $y = \frac{1}{2}x-1$.

It turns out that this is no coincidence.

> $y = mx + b$
> The graph of the equation $y = mx + b$
> is a line with slope m.

In the special case of the equation $y = b$, the value of m is zero; fortunately we already know that the graph of $y = b$ is a horizontal line and that horizontal lines have slope 0.

It turns out that the constant b in the equation $y = mx + b$ is also significant. It's time to look at b more closely.

THE INTERCEPT IN THE EQUATION

Now that we know the role of m in $y = mx + b$, let's look at a few different equations with the same m to see how the value of b affects the graph.

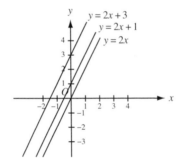

The top line is the graph of $y = 2x + 3$. The middle line is the graph of $y = 2x + 1$. The bottom line is the graph of $y = 2x$. These three lines are parallel and have the same slope. It's the value of b that distinguishes them.

In fact, the value of b is exactly the y-value at the point where the graph crosses the y-axis. Indeed, $y = 2x + 3$ crosses the y-axis at $(0, 3)$; $y = 2x + 1$ crosses at $(0, 1)$; and $y = 2x$ crosses at $(0, 0)$.

In general, $y = mx + b$ crosses the y-axis when $x = 0$, which means that y is $m(0) + b$, or b. The point $(0, b)$ is called the **y-intercept** of the equation—that's the point where the graph intersects (or *intercepts*, catches up with) the y-axis. The term *y-intercept* also refers to the number b.

KEY POINTS

Graph of y = mx + b The graph of the equation $y = mx + b$ is a line with y-intercept $(0, b)$.

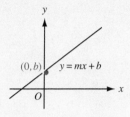

SLOPE-INTERCEPT FORM

We've figured out two important things about the graph of the equation $y = mx + b$:

1. It's a line with slope m (page 90).

2. It's a line with y-intercept b (above).

The form $y = mx + b$ is often called the **slope-intercept form** of the equation because it gives away both its slope and its y-intercept.

Slope-Intercept Form

The graph of the equation

$$y = mx + b$$

is a line of slope m through the point $(0, b)$.

EXAMPLE: Interpreting slope-intercept

What is the slope and the y-intercept of each of the lines whose equations, in slope-intercept form, are given below?

a. $y = 3x + 4$

b. $y = -\dfrac{8}{5}x$

c. $y = x - 20$

d. $y = -x + 6$

e. $y = -7$

f. $y = x$

g. $y = -x$

SOLUTION

a. Slope is 3; y-intercept is 4.

b. Slope is $-\dfrac{8}{5}$; y-intercept is 0. Keep in mind that an equation like $y = -\dfrac{8}{5}x$ is in $y = mx + b$ form with $b = 0$.

c. Slope is 1; y-intercept is –20. Keep in mind that $y = x - 20$ is the same thing as $y = 1(x) - 20$.

d. Slope is –1; y-intercept is 6. Keep in mind that $y = -x + 6$ can be rewritten as $y = -1(x) + 6$.

e. Slope is 0; y-intercept is –7. Keep in mind that $y = -7$ is the same thing as $y = 0(x) - 7$. Also, any equation in $y = b$ form is a horizontal line at height b.

f. Slope is 1; y-intercept is 0. You can think of $y = x$ as $y = 1(x) + 0$.

g. Slope is –1; y-intercept is 0. You can think of $y = -x$ as $y = -1(x) + 0$.

Slope-intercept form is good for graphing. All you have to do is look at the equation, read off the slope and the y-intercept, and draw the line.

EXAMPLE: Graph slope-intercept equation

Sketch the graph of the equation $y = \frac{1}{3}x - 2$.

SOLUTION

The equation is in slope-intercept form, so we're looking to draw a line with slope $\frac{1}{3}$ and y-intercept -2.

First, plot the y-intercept $(0, -2)$. Next, draw a line with slope $\frac{1}{3}$ through that point—move 1 unit up and 3 units to the right, and connect the dots.

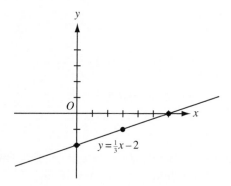

$$y = \tfrac{1}{3}x - 2$$

CONVERTING TO SLOPE-INTERCEPT FORM

One of the easiest ways of graphing a linear equation is to convert it to slope-intercept form and use the slope and the y-intercept to draw the line. That's why it's important to be able to convert a linear equation into slope-intercept form.

After slope-intercept form, the second-most popular form of a linear equation is $Ax + By = C$, called *standard form*. An $Ax + By = C$ equation with nonzero B can be converted to slope-intercept form by subtracting Ax from both sides and then dividing through by B.

EXAMPLE: Converting to slope-intercept I

Express $-3x + 4y = 5$ in slope-intercept form.

SOLUTION

To change $-3x + 4y = 5$ into slope-intercept form, first add $3x$ to both sides of the equation and then divide both sides by 4:

$$-3x + 4y = 5 \qquad \text{Rewrite the equation}$$
$$-3x + 4y + 3x = 5 + 3x \qquad \text{Add } 3x \text{ to both sides}$$
$$4y = 3x + 5 \qquad \text{Simplify}$$
$$\frac{4}{4}y = \frac{3x + 5}{4} \qquad \text{Divide both sides by 4}$$
$$y = \frac{3}{4}x + \frac{5}{4} \qquad \text{Simplify}$$

And $y = \frac{3}{4}x + \frac{5}{4}$ is in slope-intercept form.

EXAMPLE: Converting to slope-intercept II

Convert $2x - 7y = 21$ to $y = mx + b$ form.

SOLUTION

To change $2x - 7y = 21$ into $y = mx + b$ form, first subtract $2x$ from both sides and then divide both sides by -7:

$$2x - 7y = 21 \qquad \text{Rewrite the equation}$$
$$2x - 7y - 2x = 21 - 2x \qquad \text{Subtract } 2x \text{ from both sides}$$
$$-7y = -2x + 21 \qquad \text{Simplify}$$
$$\frac{-7}{-7}y = \frac{-2x + 21}{-7} \qquad \text{Divide both sides by } -7$$
$$y = \frac{-2}{-7}x + \frac{21}{-7} \qquad \text{Simplify}$$
$$y = \frac{2}{7}x - 3 \qquad \text{Simplify}$$

And $y = \frac{2}{7}x - 3$ is in $y = mx + b$ form.

There's only one minor drawback to slope-intercept form: an equation like $x = 4$ (which is in $Ax + By = C$ form with $A = 1$, $B = 0$, and $C = 4$) cannot be expressed in slope-intercept form. This makes sense for two reasons:

1. To convert $Ax + By = C$ to slope-intercept form, subtract Ax from both sides and then divide both sides by B. If B is zero, as it is in an equation like $x = 4$, you cannot do the division. So, no slope-intercept form.

2. The graph of an equation like $x = 4$ is a vertical line (page 77), and vertical lines have no slope (page 85). Since you need slope m to be able to write down slope-intercept form $y = mx + b$, you can't write down a slope-intercept form for a vertical line like $x = 4$.

Therefore, you have to deal with equations in the form $x = a$ separately, graphing them without converting to slope-intercept form first. Fortunately, this is not difficult. We already know that the graph of $x = a$ is a vertical line that passes through the points whose x-coordinate is a.

KEY POINTS

Graphing a Linear Equation

If you can convert the equation to slope-intercept form
That is, if the B in $Ax + By = C$ is not zero (most cases):

1. Convert the equation to $y = mx + b$ form.

2. Plot the y-intercept $(0, b)$.

3. Draw a line of slope m through the y-intercept.

If you cannot convert the equation to slope-intercept form
That is, if the B in $Ax + By = C$ is zero (special case):

1. Express the equation as $x = a$.

2. Plot the point $(a, 0)$. This is the **x-intercept** of the equation.

3. Draw a vertical line through $(a, 0)$.

EXAMPLE: General graphing

Graph the equation $2x + 2y = 6$.

SOLUTION

The equation has a nonzero y term, so it can be converted to slope-intercept form.

1. **Convert to slope-intercept form:**

$$2x + 2y = 6 \qquad \text{Rewrite the equation}$$
$$2x + 2y - 2x = 6 - 2x \qquad \text{Subtract } 2x \text{ from both sides}$$
$$2y = -2x + 6 \qquad \text{Simplify}$$
$$\frac{2}{2}y = \frac{-2x + 6}{2} \qquad \text{Divide by 2}$$
$$y = -x + 3 \qquad \text{Simplify}$$

2. **Plot the y-intercept:** The y-intercept of $y = -x + 3$ is $(0, 3)$; it's plotted below.

3. **Draw a line of slope *m*:** The slope of $y = -x + 3$ is -1. Here's the graph:

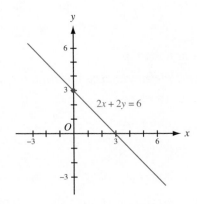

Writing Equations for Graphs

Now that we know how to convert equations into graphs, we'll discuss converting graphs into equations.

Again, it's usually easiest to use slope-intercept form. To write down the equation of a line in the coordinate plane,

1. Identify the *y*-intercept *b*.

2. Find another point and use it along with the *y*-intercept to compute the slope *m*.

3. Write the equation $y = mx + b$.

EXAMPLE: Graph to slope-intercept
Find the equation of the line below.

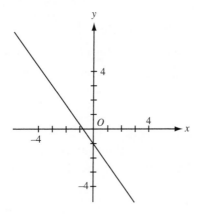

SOLUTION
Identify the *y*-intercept, find the slope, and write the equation:

1. **y-intercept:** The line crosses the *y*-axis at $(0, -1)$, so $b = -1$.

2. **Slope:** To find the slope, we need another point on the line, say $(-3, 3)$. Then $m = \dfrac{3 - (-1)}{-3 - 0} = -\dfrac{4}{3}$.

3. **Equation:** The equation, in slope-intercept form, is
$$y = -\frac{4}{3}x - 1.$$

Sometimes the graph of a line is described rather than shown, but you can still use the same methods: find the y-intercept, find the slope, write the equation.

EXAMPLE: Description to slope-intercept

Give an equation of the line of slope –3 that passes through the point (0, 7).

SOLUTION

A line with slope –3 and y-intercept 7 has slope-intercept equation $y = -3x + 7$.

EXAMPLE: Parallel description to slope-intercept

What is an equation for the line that has its y-intercept at –2 and runs parallel to $y = \frac{1}{4}x - 9$?

SOLUTION

Parallel lines have the same slope. The line $y = \frac{1}{4}x - 9$ is in slope-intercept form with slope $\frac{1}{4}$. So the line that we're interested in has slope $\frac{1}{4}$ and y-intercept –2; its equation is $y = \frac{1}{4}x - 2$.

As before, there's an exception: vertical lines don't have a y-intercept, so you can't use slope-intercept form to write down their equations. Fortunately, every vertical line has the form $x = a$.

EXAMPLE: Vertical line to equation

What is the equation of the line?

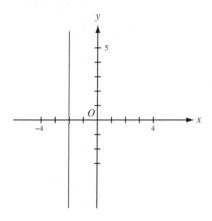

SOLUTION

The line is vertical and so has no y-intercept. (A vertical line runs parallel to the y-axis, and parallel lines never intersect.)

The general equation of a vertical line is $x = a$. This vertical line runs 2 units to the left of the x-axis, so its equation is $x = -2$.

In particular, its x-intercept is $(-2, 0)$.

POINT-SLOPE FORM

We can also find the equation of a line if we know its slope and any point on the line, not necessarily the y-intercept.

EXAMPLE: Point-slope discover
Write down an equation of the line that has slope 3 and passes through the point (1, 2).

SOLUTION
Since the slope is 3, the equation will look like $y = 3x + b$. All we have to do is figure out b. Fortunately, we know that (1, 2) is a solution—in other words, that

$$2 = 3(1) + b$$

Now we can use algebra to find b:

$$2 = 3(1) + b = 3 + b$$
$$2 - 3 = 3 + b - 3$$
$$-1 = b$$

So $b = -1$ and the equation is $y = 3x - 1$.

This method for computing b is very reliable. Alternatively, you may memorize a formula for the **point-slope form** of an equation.

> Point-Slope Formula
> The line with slope m passing through (x_0, y_0) has the form
>
> $$y - y_0 = m(x - x_0)$$

In the example above, the formula gives the equation $y - 2 = 3(x - 1)$, which is equivalent to the equation that we found:

$$y - 2 = 3(x - 1)$$
$$y - 2 = 3x - 3$$
$$y - 2 + 2 = 3x - 3 + 2$$
$$y = 3x - 1$$

The point-slope formula is a generalization of slope-intercept form. Slope-intercept form allows you to write down an equation of a line if you know its slope and its y-intercept; the point-slope formula works if you know the slope and any point on the line—which may or may not be the y-intercept.

EXAMPLE: Point-slope formula

Find an equation of the line that passes through $(7, 4)$ and $(-2, 10)$.

SOLUTION

We use the given points to find the slope:

$$m = \frac{10-4}{-2-7} = \frac{6}{-9} = -\frac{2}{3}.$$

By the point-slope formula (with point $(-2, 10)$), this line is given by the equation $y - (10) = -\frac{2}{3}(x - (-2))$, which simplifies to $y - 10 = -\frac{2}{3}(x + 2)$.

We can also use the other point to write down another form of the equation: $y - 4 = -\frac{2}{3}(x - 7)$.

These two equations are equivalent; check that both can be simplified to $y = -\frac{2}{3}x + \frac{26}{3}$. Also see the box on page 102.

OPTIONS

Bypassing the Point-Slope Formula You can bypass the point-slope form and find b directly whenever you know the slope and a point.

In the previous example, once you've found that the slope is $-\frac{2}{3}$, you know that the slope-intercept equation has the form $y = -\frac{2}{3} + b$. Now you can plug in one of the two points to find b:

$$y = -\frac{2}{3}x + b$$

$$10 = -\frac{2}{3}(-2) + b$$

$$10 = \frac{4}{3} + b$$

$$10 - \frac{4}{3} = b$$

$$\frac{26}{3} = b$$

So the equation is $y = -\frac{2}{3}x + \frac{26}{3}$.

A number of instructors prefer this method to the point-slope formula. As usual, the choice of method is yours: if you're good with formulas, use them; otherwise, reason things through.

Summary

Graphs

- A *graph* of an equation is the set of all solutions to the equation, plotted in a geometric way. Solutions to one-variable equations are plotted on the one-dimensional number line. Solutions to two-variable equations are plotted on the two-dimensional *coordinate plane*.

Distance in the plane

- The distance between point (a_1, b_1) and point (a_2, b_2) in the coordinate plane is given by

$$\sqrt{(a_2 - a_1)^2 + (b_2 - b_1)^2}$$

Slope of a line

- The *slope* of a line that passes through (x_1, y_1) and (x_2, y_2) is given by

$$m = \frac{\text{change in } y}{\text{change in } x} = \frac{y_2 - y_1}{x_2 - x_1}$$

Parallel lines

- Parallel lines have the same slope.

Eyeballing slopes

- Steep lines have slopes with big absolute value; shallow lines have slopes with small absolute value.

- Lines that go up to the right have positive slope; lines that go up to the left have negative slope.

	$|m| > 1$	$|m| < 1$
$m > 0$	$m > 1$	$0 < m < 1$
$m < 0$	$m < -1$	$-1 < m < 0$

- Horizontal lines have slope 0.
- Vertical lines have undefined slope.

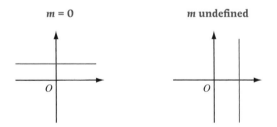

$m = 0$ m undefined

Equations of vertical and horizontal lines

- The equation of a horizontal line is $y = b$.
- The equation of a vertical line is $x = a$.

Forms of a linear equation

	Equation	Applies to	Notes
Standard form	$Ax + Bx + C$	all linear equations	
Slope-intercept form	$y = mx + b$	all but vertical lines	m is slope b is y-intercept
Point-slope form	$y - y_0 = m(x - x_0)$	all but vertical lines	m is slope (x_0, y_0) is point on line

Converting equation to graph

If the equation does not include y terms, it can be simplified to $x = a$ and the graph is a vertical line through $(a, 0)$. Otherwise,

1. Convert equation to slope-intercept form.

2. Plot y-intercept.

3. Draw line with proper slope through y-intercept.

Converting graph to equation

If the graph is a vertical line, the equation is $x = a$. Otherwise,

1. Identify y-intercept.

2. Calculate the slope.

3. Write equation in slope-intercept form.

Sample Test Questions

Answers to these questions begin on page 387.

1. Graph each equation by plotting its solutions on a number line.

    A. $2x + 7 = 4$

    B. $3(x + 4) - 2(x + 5) = 2 + x$

2. Plot the points
    $$A = (5, 0), \quad B = \left(0, -\frac{5}{2}\right), \quad C = (-4, 1), \quad D = (2, 7)$$
    in the coordinate plane. Which of them are in the upper half-plane? Which are in quadrant III?

3. What's the distance between each pair of points in question 2?

    A. B and the origin

    B. C and the origin

    C. D and C

4. What is the slope of the line that connects each pair of points below from question 2?

    A. C and the origin

    B. D and C

    C. A and B

    D. B and the origin

5. A square in the coordinate plane has three of its vertices at $(6, 0)$, $(-1, -1)$, and $(3, -4)$. What are the coordinates of the fourth vertex? (A square has four right angles and four equal sides; opposite sides are parallel.)

6. Which of these equations are linear?

    A. $3x(1 - x) = y - 9$

    B. $y = x$

    C. $3(1 - x) + 4(x + y) = y$

    D. $x = 3y - 5$

7. Make a table of values for the equation $x = 3y - 5$ and use it to plot the graph of the equation. Use the graph to determine the value of y when $x = 10$.

8. Graph each of these equations in the coordinate plane.

   A. $y = x + 4$

   B. $x = 6$

   C. $2y = -6$

   D. $y = -2x - 5$

   E. $2x - 3y = 1$

9. Give an equation for each of these lines.

   A.

   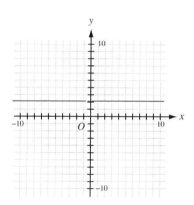

   B.

   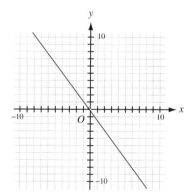

C.

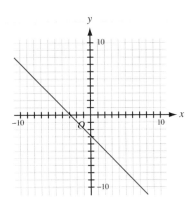

D.

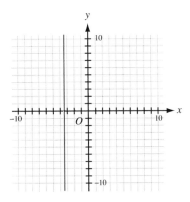

E.

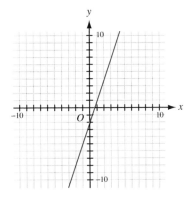

10. A line in the plane passes through the points (–7, 9) and (a, –2). If the slope of the line is $-\dfrac{1}{2}$, what is the value of a?

11.

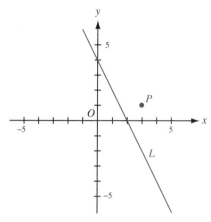

The graph shows a line L and a point P. What is an equation of the line parallel to L that passes through P?

12. Give an equation, in standard form, for the line that passes through the point (2, 3) and runs parallel to the line $5x - 4y = 7$.

13. Give an equation for the line with slope 0 and y-intercept (0, 0).

14. A *parallelogram* is a four-sided figure whose opposite sides are parallel. Do these four points form a parallelogram in the plane

(6, 8),   (–1, –6),   (–5, –3),   (1, 9)?

15. Helmut is signing up volunteers to help organize next year's Oktoberfest celebration. When he started at 11:30 A.M. one Monday, he already had 4 volunteers that he'd found over the weekend. By 3:30 P.M., he had signed up a total of 18 volunteers.

    A. Find the average rate, in new volunteers per hour, of Helmut's recruitment efforts that Monday.

    B. Write an equation that expresses the number of Helmut's Oktoberfest volunteers n in terms of the number of hours t after 11:30 A.M. (Assume that this equation is linear, that is, that Helmut's recruitment rate was pretty much the same during the day.)

    C. If Helmut keeps it up, how many volunteers can he expect to have by 7 P.M. ($t = 7.5$)?

16. Elizabeth stopped cutting her hair in July of 2005. She's noticed that the length of her hair is approximately given by the formula

    $$h = 1.2t + 7.5,$$

    where t is the number of months since she stopped cutting her hair in July and h is hair length in inches.

    A. About how long was Elizabeth's hair in July ($t = 0$)? In November?

    B. If Elizabeth continues not to cut her hair, when can she expect it to reach 30 inches?

    C. Graph the equation that describes Elizabeth's hair growth. Use a horizontal t-axis and a vertical h-axis.

17. A shuttle bus drives tourists from a hotel to Disney World, 24 miles away. The figure below is the graph of the mile distance *d* of the bus from Disney World as the trip progresses in minutes *t*.

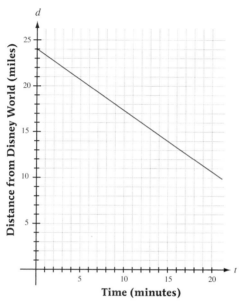

A. How far is the bus from Disney World after 15 minutes of travel?

B. How long does it take for the bus to travel 4 miles?

C. What is the *d*-intercept of the graph? What real-world quantity does it represent?

D. What is the slope of the line in the graph? What real-world quantity does the slope represent?

E. Use the graph to write down an equation that expresses the distance from Disney World *d* in terms of the time *t*.

F. If the bus always goes at the same speed, how long does the trip from the hotel to Disney World take?

Systems of Linear Equations

3

Overview

Let's say you urgently have to get to Raleigh, North Carolina, from New York City. A bus is leaving in a few minutes. Just before you get on board, you learn that there's another bus; it's not leaving for another few hours, but it travels much faster. Which bus do you take?

Of course, the answer is, it depends—on the relative speeds of the two buses, on the wait time between the two departures, on the distance to Raleigh, perhaps on the difference in ticket price. But even if you knew all these pieces of information, how do you decide?

One extremely reliable way is to put together a *system of equations*. One equation represents the first bus; the other represents the second. At some point, if they travel far enough along the same route, the faster bus will overtake the slower. The question for you is, will this happen before or after the slower bus reaches Raleigh, North Carolina? At the moment of overtaking, the two buses will be in the same place at the same time—their equations intersect. All you have to do is figure out where this intersection takes place.

That's how a system of equation works: two changing quantities (in this case, time and space) are related in several different ways (through the motion of the two buses). A system of equations is a framework for organizing information that's too complicated to keep track of intuitively for a busy human brain.

Linear equations are the simplest kind of equations; they're useful both in their own right and as cruder approximations to more-complicated unsolvable equations. In this chapter, we look for common solutions to systems of linear equations in two variables, both graphically, using methods developed in Chapter 2, and algebraically, using methods based on those developed in Chapter 1.

Simultaneous Equations

Two or more equations considered together are called **simultaneous equations**, or a **system of equations**. An opening brace ({) is sometimes used to indicate that two or more equations should be considered together, like so:

$$\begin{cases} x^2 + y^2 = 25 \\ xy = 12 \end{cases}$$

A solution to a system of equations in two variables is an ordered pair that satisfies *both* equations. Try this example on your own.

EXAMPLE:

Consider the simultaneous equations

$$\begin{cases} x^2 + y^2 = 25 \\ xy = 12 \end{cases}$$

Which of these points are solutions to this system of equations?

a. $(4, -3)$

b. $(2, 6)$

c. $(-3, -4)$

d. $(2, 1)$

SOLUTION

Take each point and check each equation.

a. Plugging $(4, -3)$ into the first equation gives $(4)^2 + (-3)^2 = 25$, which is a true statement. Plugging into the second equation gives $(4)(-3) \overset{?}{=} 12$, which is a false statement. This ordered pair is a solution to the first equation but not to the second, so it is *not* a solution to the system.

b. Plugging (2, 6) into the first equation gives $(2)^2 + (6)^2 \overset{?}{=} 25$, which is a false statement. At this point, we already know that (2, 6) is not a solution to the system of equations. For the record, it is a solution to the second equation, since $(2)(6) = 12$ is a true statement.

c. Plugging (–3, –4) into the first equation gives $(-3)^2 + (-4)^2 = 25$, which is a true statement. Plugging into the second equation gives $(-3)(-4) = 12$, which is also a true statement. So (–3, –4) is a solution to the system of equations.

d. Plugging (2, 1) into the first equation gives $(2)^2 + (1)^2 \overset{?}{=} 25$, which is a false statement. At this point, we can conclude that (2, 1) is not a solution to the system of equations. For the record, (2, 1) is also not a solution to the second equation, since $(2)(1) \overset{?}{=} 12$ is a false statement.

For the remainder of this chapter, we'll be looking at systems of *linear* equations, though most of the methods that we'll discuss also work for more complicated equations.

Graphical Solution Method

The points on the graph of an equation are all solutions to that equation. So if we graph two equations on the same set of axes, the points that are on *both* graphs will be solutions to *both* equations. In other words, we can find solutions to a system of two equations by graphing both equations and finding points common to both graphs. These common points are called the **intersection** of the two graphs.

Let's see this in action.

EXAMPLE: Solving by graphing I
Find the solutions to the system

$$\begin{cases} x + y = 6 \\ 2y - x = 18 \end{cases}$$

by graphing both equations.

SOLUTION
Each of these is a linear equation, so the graph will be a straight line. One quick way to graph a linear equation is to plot two points and then draw the line connecting them. The x-intercept and the y-intercept are two great candidates—they're easy to find by plugging in zero for each variable in turn.

The equation $x + y = 6$ passes through $(6, 0)$ and $(0, 6)$.
The equation $2y - x = 18$ passes through $(-18, 0)$ and $(0, 9)$.
Plot the points and draw the graphs:

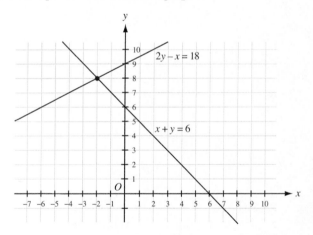

The two lines intersect at $(-2, 8)$. This is the only ordered pair in the coordinate plane that's a solution to both equations.

Let's check the answer: $(-2) + (8) = 6$, so $(-2, 8)$ is a solution to the first equation; $2(8) - (-2) = 18$, so $(-2, 8)$ works in the second equation as well.

Try another example by yourself before reading the solution.

EXAMPLE: Solving by graphing II

Graph each equation to find a simultaneous solution to

$$\begin{cases} 7x = y - 25 \\ y = -3 \end{cases}$$

SOLUTION

First, convert both equations to slope-intercept form. The first equation becomes $y = 7x + 25$. The second equation is already in slope-intercept form.

Next, read off the slopes and the y-intercepts. The first graph has slope 7 and crosses the y-axis at $(0, 25)$. The second graph is a horizontal line at -3.

Then graph:

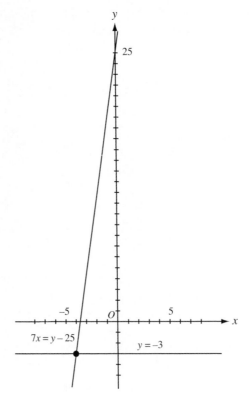

Finally, find the common solution by reading off the coordinates of the point of intersection. The two lines intersect at (–4, –3), so that's the common solution.

Check: $7(-4) = -3 - 25$, so the point works in the first equation. Since the y-coordinate is –3, the point also works in the second equation.

SPECIAL CASES

Most of the time, two lines will intersect at exactly one point, so a system of two equations in two variables will have exactly one solution. But there are special cases with no solutions and special cases with many solutions.

Special Case: Parallel Lines

EXAMPLE: *Solving by graphing: parallel lines*
Solve the following system of equations by graphing:

$$\begin{cases} y = -2x + 6 \\ 2y + 4x = 3 \end{cases}$$

SOLUTION
The first equation is already in slope-intercept form: its graph
will have slope –2 and y-intercept (0, 6).

The second equation is not in slope-intercept form, so convert:

$$2y + 4x = 3$$
$$2y + 4x - 4x = 3 - 4x$$
$$2y = -4x + 3$$
$$\frac{2y}{2} = -\frac{4}{2}x + \frac{3}{2}$$
$$y = -2x + \frac{3}{2}$$

So the graph of the second equation has slope –2 and y-intercept $\left(0, \frac{3}{2}\right)$.

Graph both equations on the same set of axes:

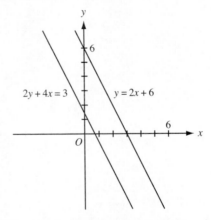

The graphs of the two equations are parallel, and parallel lines
never intersect. So there are no ordered pairs that are solutions to
both equations. This is an *inconsistent* system. It has no solutions.

KEY POINTS

Consistent and Inconsistent Systems A **consistent** system of equations has at least one solution. Each equation gives information that is *consistent* with the information given by the other equation; there are no contradictions.

An **inconsistent** system of equations describes a scenario that can't possibly happen: for example, $x + y = 0$ and $x + y = 1$ simultaneously. Each equation gives information that's *inconsistent* with the other one's.

Special Case: Same Line

EXAMPLE: Solving by graphing: same line

Solve the following system of equations by graphing:

$$\begin{cases} y = x - 2 \\ 4y + 6 = 3x + y \end{cases}$$

The first equation is already in slope-intercept form; the graph has slope 1 and y-intercept -2.

The second equation needs to be converted:

$$4y + 6 = 3x + y$$
$$4y + 6 - y - 6 = 3x + y - y - 6$$
$$3y = 3x - 6$$
$$\frac{3y}{3} = \frac{3}{3}x - \frac{6}{3}$$
$$y = x - 2$$

It turns out that the two equations are equivalent. They have the same solutions and identical graphs:

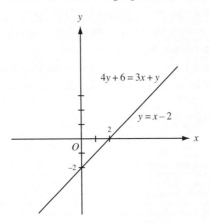

The intersection of their graphs is the whole line. So this system of equations has infinitely many solutions: namely, all the points that are on the line given by $y = x - 2$. A system like this, where two equations in two variables have the same solution set, is called *dependent*.

KEY POINTS

System of two linear equations in two variables There are three scenarios for a system of two linear equations in two variables:

- **Exactly one solution:** The graph is the intersection of two distinct nonparallel lines—that is, a point.

- **No solutions:** The graph is the intersection of two distinct parallel lines—that is, nothing.

- **An infinite number of solutions:** The graph is the intersection of a line with itself—that is, the whole line.

	Graphics	Solutions	Consistent?	Independent?
Different slopes	intersecting lines	one	consistent	independent
Same slope, diff. intercepts	parallel lines	zero	inconsistent	independent
Same slope, same intercepts	same line	infinite	consistent	dependent

Algebraic Solution Methods

Graphing a system of equations makes it easy to see what kind of system you're dealing with. But when you're searching for solutions, it may be difficult or impractical to find the exact answer just by looking at a graph. Fortunately, exact solutions can be obtained by algebraic methods.

There are two popular methods. Both methods reduce two two-variable equations to a one-variable equation that can be solved easily.

- **The substitution method:** You solve one of the equations to express one variable in terms of the other. That is, you massage one of the equations into an "$x =$" or "$y =$" form. Then you plug the expression for x or y into the other equation to get a one-variable equation. This is a simple yet powerful technique that is important to master because it's the basis of most algebraic techniques for solving a wide range of equations.

- **The adding-equations method** (sometimes known as the *linear combination method* or the *elimination method*): This works with linear equations only. It involves adding multiples of the two equations in a clever way so as to eliminate one of the variables. It's generally cleaner and more elegant but requires more planning than substitution. If you're unlikely to study matrices or linear algebra in the future, you can probably skip this method.

As usual, neither method is better or more correct.

SUBSTITUTION METHOD

Let's dig right in.

EXAMPLE: Easier substitution

Solve the following system of equations:

$$\begin{cases} x = 8 - 4y \\ 2x + 5y = 13 \end{cases}$$

SOLUTION

The first equation gives us an expression for x in terms of y. We can plug this expression into the second equation:

$$2x + 5y = 13$$
$$2(8 - 4y) + 5y = 13$$

Now we have an equation in one variable, which is easier to solve.

$$2(8 - 4y) + 5y = 13$$
$$16 - 8y + 5y = 13$$
$$16 - 3y = 13$$
$$16 - 3y - 16 = 13 - 16$$
$$-3y = -3$$
$$\frac{-3y}{-3} = \frac{-3}{-3}$$
$$y = 1$$

So the solution to this system of equations has a y-value of 1. To find x, we plug this y-value into the expression for x in terms of y:

$$x = 8 - 4y$$
$$x = 8 - 4(1) = 4$$

So the solution is $(4, 1)$.

Check the second equation just to be sure (we used the first equation to get x after we knew y, so checking the first equation is unlikely to be useful): $2(4) + 5(1) = 13$ is true, so $(4, 1)$ is the solution.

That's basically all there is to the substitution technique. It won't always be so easy, of course. Sometimes you'll have to do some manipulation to get an expression for one variable in terms of the other.

EXAMPLE: Harder substitution

Use substitution to solve the following system of equations:

$$\begin{cases} 7x + 3y = -1 \\ 9x - 5y = 12 \end{cases}$$

SOLUTION

First we use one equation to express one variable in terms of the other. You target either variable in either equation. Sometimes you can be clever and make choices that minimize your work, but in this case it doesn't much matter. So let's express y in terms of x in the first equation:

$$7x + 3y = -1$$
$$7x + 3y - 7x = -1 - 7x$$
$$3y = -1 - 7x$$
$$y = \frac{-1 - 7x}{3}$$

Now, substitute this expression in for y in the second equation:

$$9x - 5y = 12$$
$$9x - 5\left(\frac{-1 - 7x}{3}\right) = 12$$
$$9x - 5\left(\frac{-1}{3}\right) - 5\left(\frac{-7x}{3}\right) = 12$$
$$9x + \frac{5}{3} + \frac{35}{3}x = 12$$

At this point, we can collect like terms and solve for x. But the numbers that we're dealing with will be nicer if we can clear out the denominators, so let's multiply both sides of the equation by 3:

$$3 \cdot 9x + 3 \cdot \frac{5}{3} + 3 \cdot \frac{35}{3}x = 3 \cdot 12$$
$$27x + 5 + 35x = 36$$
$$62x + 5 = 36$$

Collect like terms and solve for x:

$$62x + 5 - 5 = 36 - 5$$
$$62x = 31$$
$$x = \frac{31}{62} = \frac{1}{2}$$

Finally, plug $x = \frac{1}{2}$ into the expression for y in terms of x:

$$y = \frac{-1 - 7x}{3}$$

$$y = \frac{-1 - 7\left(\frac{1}{2}\right)}{3}$$

$$y = \frac{-\frac{9}{2}}{3}$$

$$y = -\frac{3}{2}$$

So $\left(\frac{1}{2}, -\frac{3}{2}\right)$ is the solution.

The math was messy here; check the solution in both equations: $7\left(\frac{1}{2}\right) + 3\left(-\frac{3}{2}\right) = -\frac{2}{2} = -1$, so it works in the first equation; $9\left(\frac{1}{2}\right) - 5\left(-\frac{3}{2}\right) = \frac{24}{2} = 12$, so it works in the second also.

The messiness of this problem is exactly the kind of thing that the adding-equations method cleans up. We'll get to it a little later, after we solve a few different kinds of systems with the substitution method.

Smart choices, about which equation to manipulate first and which variable to solve for, can often cut down on the work.

EXAMPLE: Medium substitution

Find all the solutions to the system of equations:

$$\begin{cases} 10x + 7y = 49 \\ 10y - x = 70 \end{cases}$$

SOLUTION

In this system of equations, the easiest variable to isolate is the x in the second equation. Its coefficient is -1, so it won't create any denominators.

$$10y - x = 70$$
$$10y - x - 70 + x = 70 - 70 + x$$
$$10y - 70 = x$$

Now use the other equation:

$$10x + 7y = 49$$
$$10(10y - 70) + 7y = 49$$
$$100y - 700 + 7y = 49$$
$$107y - 700 + 700 = 49 + 700$$
$$107y = 749$$
$$y = \frac{749}{107} = 7$$

Finally, solve for x:

$$x = 10y - 70$$
$$x = 10(7) - 70 = 0$$

So $(0, 7)$ is the solution. Check on your own that it works in both equations.

Special Cases

We've discussed some special cases of simultaneous equations geometrically; let's see what they look like algebraically.

EXAMPLE: No solution substitution

Solve the following system of equations:

$$\begin{cases} 4x - 6y = 7 \\ 9y - 6x = -5 \end{cases}$$

SOLUTION

Approach this problem like any other. Use one of the equations to solve for one of the variables—say, for x in the first equation:

$$4x - 6y = 7$$

$$4x = 6y + 7$$

$$x = \frac{6}{4}y + \frac{7}{4}$$

$$x = \frac{3}{2}y + \frac{7}{4}$$

Substitute for x in the second equation:

$$9y - 6\left(\frac{3}{2}y + \frac{7}{4}\right) = -5$$

$$9y - 6\left(\frac{3}{2}y\right) - 6\left(\frac{7}{4}\right) = -5$$

$$9y - 9y - \frac{21}{2} = -5$$

$$-\frac{21}{2} = -5$$

We have a false statement on our hands. This is a sure indication that this system of equations is *inconsistent* and has no solution.

Geometrically, this example is the two-parallel-line case: same slope, different *y*-intercept, so no solution. Algebraically, such a scenario ends in a false statement like $-\frac{21}{2} = -5$.

EXAMPLE: *Many solutions substitution*
Solve the following system of equations:

$$\begin{cases} x + 2y = 3 \\ -2x - 4y = -6 \end{cases}$$

SOLUTION
The coefficient of *x* in the first equation is 1, so let's solve for it:

$$x + 2y = 3$$
$$x = 3 - 2y$$

Plug this into the second equation:

$$-2(3 - 2y) - 4y = -6$$
$$-6 + 4y - 4y = -6$$
$$-6 = -6$$

We have a statement that's always true. This is a signal that the equations in the system are *dependent*; there are infinitely many solutions. The equation $x = 3 - 2y$ that we derived is still true; any ordered pair that satisfies it is a solution to this system of equations.

We can also say that the solutions to this system of equations are all ordered pairs of the form $(x, y) = (3 - 2y, y)$.

Geometrically, this example is a case of the intersection of a line with itself: same slope, same *y*-intercept, same line. Algebraically, such a scenario ends in an always-true statement, like –6 = –6.

ADDING-EQUATIONS METHOD

As mentioned before, the adding-equations method is not indispensable. However, many textbooks teach it, and many instructors use it. And it really does streamline many a simultaneous-equation solution.

EXAMPLE: Easier adding

Solve the following system of equations:

$$\begin{cases} x + 2y = 11 \\ x - 2y = 3 \end{cases}$$

SOLUTION

This problem is easy enough with substitution, but with adding equations it becomes *even easier*.

Let's see what happens when we add these equations together, left side to left side, right side to right side:

$$\begin{array}{r} x \quad\quad + 2y = 11 \\ + \quad x \quad\quad - 2y = 3 \\ \hline x + x + 2y - 2y = 11 + 3 \end{array}$$

or

$$2x = 14$$

The y terms cancel. Now it's easy to solve for x. If $2x = 14$, then $x = 7$.

At this point, we can substitute this value of x back into one of the equations to solve for y:

$$\begin{aligned} x + 2y &= 11 \\ 7 + 2y &= 11 \\ 2y &= 11 - 7 = 4 \\ y &= \frac{4}{2} = 2 \end{aligned}$$

The solution we get is $(7, 2)$. Check it in both equations to make sure that it works.

That's the gist of the adding-equations method: adding the two equations so as to eliminate one of the variables.

The previous example had some special features that made the adding-equations method particularly lucrative. The coefficients on the y terms in the two equations were additive inverses: Same magnitude, opposite signs. That's why the ys disappeared when the equations were added.

When the coefficients aren't set up to cancel, one or both of the equations can be multiplied by a cleverly chosen factor.

EXAMPLE: Double adding

Solve the following system of equations:

$$\begin{cases} 2y - 3x = 7 \\ 5x = 4y - 12 \end{cases}$$

SOLUTION

Before adding equations, you must rearrange the equations so that the x terms, the y terms, and the constant terms line up. It's conventional (but not necessary) to keep the variables on one side, xs first and ys second, and the constant on the other, so that each equation is in $Ax + By = C$ form. The rearranged system looks like this:

$$\begin{cases} -3x + 2y = 7 \\ 5x - 4y = -12 \end{cases}$$

Now we look for coefficients that are similar, factorization-wise. For example, the coefficient on y is 2 in the first equation and -4 in the second equation. If we multiply the first equation by 2, we will be able to cancel the ys:

$$\begin{cases} -6x + 4y = 14 \\ 5x - 4y = -12 \end{cases}$$

Now, add the two equations together:

$$\begin{array}{r} -6x + 4y = 14 \\ + \quad 5x - 4y = -12 \\ \hline -x \quad\quad = 2 \end{array}$$

Since $-x = 2$, we must have $x = -2$.

At this point, we again have two options. We can

1. Substitute the value of x into one of the equations and solve for y or

2. Multiply the original equations and combine them to get the xs to cancel out to solve for y.

Substitution is easier in this case, but since we're focusing on the adding-equations method, we'll demonstrate the second option.

The x coefficients in the first and second equations are -3 and 5, respectively. If we multiply the first equation by 5 and the second equation by 3, we'll be able to eliminate xs when we add the two equations together:

$$\begin{cases} 5 \left[-3x + 2y = 7 \right. \\ 3 \left[\ \ 5x - 4y = -12 \right. \end{cases} \text{ becomes } \begin{cases} -15x + 10y = 35 \\ \ \ 15x - 12y = -36 \end{cases}$$

Now add:

$$\begin{array}{r} -15x + 10y = 35 \\ + \ \ 15x - 12y = -36 \\ \hline -2y = -1 \end{array}$$

Since $-2y = -1$, $y = \dfrac{1}{2}$ and the solution is $\left(-2, \dfrac{1}{2} \right)$.

Special Cases

The adding-equations method works well in special cases where there are no solutions or infinitely many solutions. If you're attentive, you'll be able to weed them out even before adding or subtracting.

EXAMPLE: No solution adding

Solve the following system of equations:

$$\begin{cases} 7x - 6y = 20 \\ \quad 18y = 44 + 21x \end{cases}$$

SOLUTION

First, rearrange the equations:

$$\begin{cases} \quad 7x - 6y = 20 \\ -21x + 18y = 44 \end{cases}$$

Next, scale the equations to ensure that one of the variables will cancel. The coefficient on x is 7 in the first equation and –21 in the second equation, so if we multiply the first equation by 3 the xs will cancel when the equations are added:

$$\begin{array}{r} 21x - 18y = 60 \\ + \ -21x + 18y = 44 \\ \hline 0x + 0y = 104 \end{array}$$

The xs canceled, but so did the ys, leaving 0 = 104. This is a false statement, which means that the system of equations is inconsistent and has no solution. These equations describe two parallel lines in the coordinate plane, and parallel lines don't intersect.

EXAMPLE: Many solutions adding

Solve the following system of equations:

$$\begin{cases} 4y - 6x = 10 \\ 6y = 15 + 9x \end{cases}$$

SOLUTION

First, rearrange the equations.

$$\begin{cases} -6x + 4y = 10 \\ -9x + 6y = 15 \end{cases}$$

Next, multiply one or both equations by factors to make one of the variables drop out when the equations are added. The y coefficient in the first equation is 4; in the second, 6. The LCM of 4 and 6 is 12. So multiplying the first equation by 3 and the second equation by –2 should do the trick, leaving a $12y$ in the first equation and a $-12y$ in the second, so that the ys will cancel when the equations are added:

$$\begin{cases} 3 \left[-6x + 4y = 10 \right] \\ -2 \left[-9x + 6y = 15 \right] \end{cases} = \begin{cases} -18x + 12y = 30 \\ 18x - 12y = -30 \end{cases}$$

Now, add the two equations together:

$$\begin{array}{r} -18x + 12y = 30 \\ + \quad 18x - 12y = -30 \\ \hline 0x + \quad y = \quad 0 \end{array}$$

Since the sum of the two equations is $0 = 0$, a statement that's always true, the two equations are dependent and have an infinite number of solutions. These equations describe the same line in the coordinate plane.

For completeness, you could solve for one variable in terms of the other (in either equation) to give a complete description of the solutions. We'll use the original first equation, which has smaller coefficients, and solve for y in terms of x:

$$-6x + 4y = 10$$

$$4y = 10 + 6x$$

$$y = \frac{5}{2} + \frac{3}{2}x$$

So the solutions for this system of equations are all the ordered pairs in the form $(x, y) = \left(x, \frac{5}{2} + \frac{3}{2}x \right)$.

OPTIONS

Subtracting Equations If you're extremely careful with minus signs, you could try *subtracting* equations when there's a variable that has the same coefficient in both equations. Here's an example that begs for subtraction:

EXAMPLE: Subtracting
Solve the system of equations:

$$\begin{cases} 3x + 2y = 7 \\ 3x - 4y = 13 \end{cases}$$

SOLUTION
Since the *x* coefficient on both equations is 3, the *x*s will cancel if you subtract one equation from the other. Take a look:

$$\begin{array}{r} 3x \quad\quad + 2y = 7 \\ - \quad 3x \quad\quad - 4y = 13 \\ \hline 3x - 3x + 2y - (-4y) = 7 - 13 \end{array}$$

which simplifies to $6y = -6$, or $y = -1$. Substituting $y = -1$ back into either equation gives $x = 3$, so the solution is $(3, -1)$.

The example worked out well, but it can be difficult to keep track of all the minus signs. This is especially true when you're subtracting a term that already has a minus sign, as with the *y*s in the example.

For most people it's safer to multiply one of the equations by -1 and then add the two equations together, which also ends up eliminating the *x* terms. That being said, subtracting equations does cut out an extra step. In short, subtract at your own risk—and always check the answer.

Word Problems

Try to work these problems out on your own before reading through the solutions. Use whatever method of solution you prefer.

EXAMPLE: Word problem: counting

A circus act involved chickens and zebras, 13 animals in all. Emily sat in the front row and counted a total of 34 legs running around. How many zebras were there? How many chickens?

SOLUTION

1. **Restate what's known:** The problem tells us two pieces of information:

$$\text{chickens} + \text{zebras} = 13$$
$$\text{chicken legs} + \text{zebra legs} = 34$$

   We can also rely on some common knowledge: namely, that each chicken has 2 legs and each zebra has 4 legs.

2. **Choose variables:** Since we're asked to find the number of zebras and the number of chickens, let

$$c = \text{number of chickens}$$
$$z = \text{number of zebras}$$

3. **Translate into math:** First,

$$c + z = 13$$

   On to the legs. Since each chicken has 2 legs, there are $2c$ chicken legs. Each zebra has 4 legs, so there are $4z$ zebra legs. So the second equation is

$$2c + 4z = 34$$

   So solving the problem means solving the system of equations

$$\begin{cases} c + z = 13 \\ 2c + 4z = 34 \end{cases}$$

4. **Solve the system of equations:** We'll do it by substitution. The first equation is simpler; we'll use it to solve for z in terms of c:

$$c + z = 13 \qquad \text{Write the equation}$$
$$c + z - c = 13 - c \qquad \text{Subtract } c \text{ from both sides}$$
$$z = 13 - c \qquad \text{Simplify}$$

Now, plug this expression into the second equation and solve:

$$2c + 4z = 34 \qquad \text{Write the equation}$$
$$2c + 4(13 - c) = 34 \qquad \text{Substitute } 13 - c \text{ for } z$$
$$2c + 52 - 4c = 34 \qquad \text{Distribute}$$
$$-2c + 52 = 34 \qquad \text{Simplify}$$
$$-2c + 52 - 52 = 34 - 52 \qquad \text{Subtract 52 from both sides}$$
$$-2c = -18 \qquad \text{Simplify}$$
$$\frac{-2c}{-2} = \frac{-18}{-2} \qquad \text{Divide both sides by } -2$$
$$c = 9 \qquad \text{Simplify}$$

Finally, plug $c = 9$ into $z = 13 - c$ to find that $z = 4$.

5. **Translate back and check:** There were 9 chickens and 4 zebras, which makes 13 animals and $2(9) + 4(4) = 34$ legs.

EXAMPLE: Word problem: mixtures

Mr. Twister's favorite snack is a mix of cashews and prunes that's exactly $\frac{1}{3}$ cashew and $\frac{2}{3}$ prune, by weight. He manages to find a cashew-prune mix in a store—but alas, it's only $\frac{1}{5}$ cashew and $\frac{4}{5}$ prune. Fortunately he has another cashew-prune mix left over from a party; this one's half cashew, half prune. How many ounces of the first mix and how many of the second should he use to get exactly 30 ounces of his favorite snack?

SOLUTION

1. **Restate what's known:** The problem gives us four pieces of information:

   the store mix is $\frac{1}{5}$ cashew and $\frac{4}{5}$ prune

   the party mix is $\frac{1}{2}$ cashew and $\frac{1}{2}$ prune

   the perfect mix is $\frac{1}{3}$ cashew and $\frac{2}{3}$ prune

   the perfect mix is 30 ounces

2. **Choose variables:** We're asked to find how much store mix and how much party mix need to be combined to make the ideal mix. So let

   s = amount of store mix and
   p = amount of party mix

3. **Translate into math:** This is a classic mixture problem, and the best way to deal with it is to make a classic mixture table. We'll only keep track of the cashews in each mixture, since you can always find the fraction or amount of prunes by subtracting from the total.

	Store mix	Party mix	Perfect mix
Amount of mixture	s	p	30 oz
Fraction of cashews	$\frac{1}{5}$	$\frac{1}{2}$	$\frac{1}{3}$
Amount of cashews			

All three columns form multiplication equations. So you can fill in the bottom row:

	Store mix	Party mix	Perfect mix
Amount of mixture	s	p	30 oz
Fraction of cashews	$\frac{1}{5}$	$\frac{1}{2}$	$\frac{1}{3}$
Amount of cashews	$\frac{1}{5}s$	$\frac{1}{2}p$	$\frac{1}{3}(30)$ oz

The *first* and *third* rows also form addition equations. So we can write them down:

$$s + p = 30$$
$$\frac{1}{5}s + \frac{1}{2}p = \frac{1}{3}(30)$$

(Watch out! The second row of a mixture table gives a rate, or a percentage, or a fraction. You cannot add across this row. You may use only the first and the third rows to write addition equations.)

4. **Solve the system of equations:** We'll do this one using the adding-equations method.

Beforehand, simplify the right side of the second equation to get:

$$\frac{1}{5}s + \frac{1}{2}p = \frac{1}{3}(30) = 10$$

and multiply both sides by $5 \cdot 2 = 10$, clearing denominators, to get:

$$2s + 5p = 100$$

We'll multiply the equation $s + p = 30$ by –2:

$$-2s - 2p = -60$$

so that the s terms will cancel when we add the two equations together:

$$
\begin{array}{r}
-\,2s - 2p = -60 \\
+\ \ 2s + 5p = 100 \\
\hline
3p = \ \ 40
\end{array}
$$

Since $3p = 40$, we know that $p = \dfrac{40}{3}$, or $13\frac{1}{3}$, and $s = 30 - p = 16\frac{2}{3}$.

5. **Translate back and check:** So Mr. Twister needs to use $16\frac{2}{3}$ ounces of the store mix and $13\frac{1}{3}$ ounces of the party mix to make his ideal snack.

Check to make sure that the cashew content works out: the new mixture will have

$$\frac{1}{5}\left(16\frac{2}{3}\right) + \frac{1}{2}\left(13\frac{1}{3}\right) = \frac{1}{5}\left(\frac{50}{3}\right) + \frac{1}{2}\left(\frac{40}{3}\right) = \frac{10}{3} + \frac{20}{3} = 10$$

ounces of cashews, which is exactly $\frac{1}{3}$ of the 30 desired ounces of mix.

Summary

A system of two linear equations

$$\begin{cases} Ax + By = E \\ Cx + Dy = F \end{cases}$$

must be one of three types:

1. One solution

- **Shape of graph:** two intersecting lines
- **Set of solutions:** one point (the point of intersection)
- **Number of solutions:** one
- **Algebraically reduces to:** a single value for x and a single value for y
- **Consistent or inconsistent:** consistent
- **Dependent or independent:** independent

2. No solutions

- **Shape of graph:** two parallel lines
- **Set of solutions:** empty
- **Number of solutions:** zero
- **Algebraically reduces to:** a false statement, like $0 = -2$
- **Consistent or inconsistent:** inconsistent
- **Dependent or independent:** independent

3. Many solutions

- **Shape of graph:** one line (given by both equations)
- **Set of solutions:** one line (that same one)
- **Number of solutions:** infinitely many
- **Algebraically reduces to:** a true statement, like $0 = 0$
- **Consistent or inconsistent:** consistent
- **Dependent or independent:** dependent

Sample Test Questions

Answers to these questions begin on page 392.

1. Which of these ordered pairs are solutions to the simultaneous equations

$$\begin{cases} 11x + 4y = 3 \\ 19x + 7y = 5 \end{cases}$$

   A. $(-3, 9)$

   B. $(1, -2)$

   C. $\left(0, \dfrac{5}{7}\right)$

2. Graph this system of equations to find all common solutions:

$$\begin{cases} y = 2x - 3 \\ y = -\dfrac{1}{2}x + 2 \end{cases}$$

3. Solve this system of equation by graphing:

$$\begin{cases} y - 3 = 2(x - 1) \\ 2x - y + 6 = 0 \end{cases}$$

4. Solve this system of equations using substitution:

$$\begin{cases} -3x + 8y = 18 \\ 2x - 5y = -11 \end{cases}$$

5. Use substitution to find all the pairs (x, y) that satisfy

$$\begin{cases} \dfrac{1}{3}x - \dfrac{4}{3}y = -3 \\ -1.5x + 6y = 13.5 \end{cases}$$

6. Use the adding-equations method to solve this system of linear equations:

$$\begin{cases} 6x + 5y = 8 \\ 3x - 2y = -14 \end{cases}$$

7. Solve the system of equations:

$$\begin{cases} 4x - 3y = 0 \\ -8x + 9y = 2 \end{cases}$$

8. What are all solutions to the system of equations

$$\begin{cases} 2.5x - 3y = 5.5 \\ 4x - 3y = -2 \end{cases}$$

9. Rob and Sam enter a pie-eating contest in which each contestant eats apple pies from the beginning of the contest until noon and then switches to pecan pies until the end of the contest. Rob can eat 3 apple pies or 5 pecan pies an hour, whereas Sam can eat 4 apple pies or 7 pecan pies an hour. If Rob eats 11 pies during the contest and Sam eats 15 pies, what time does the contest start? How long does it last?

10. At the end of the day on a Friday in 1992, the Curious Liquids Coffeehouse tip jar held only dimes and quarters, 90 coins in all. Each of the three employees went home with $4.85 in tips for the day. How many quarters and how many dimes did the Curious Liquids Coffeehouse tip jar contain that Friday night?

Exponents, Roots, Radicals

4

Overview

You're probably already familiar with *exponents*—those little numbers that sit to the upper right of an expression and tell you how many times it is multiplied by itself. For example, *four squared* is 4^2 or $4 \cdot 4$, which comes out to 16. And you've probably already seen some roots and radicals, especially square roots. Taking a square root is the reverse of squaring: if you start with 16, you'll get back 4.

After some review, we'll move on to new topics. First, negative exponents, which appear in expressions like 5^{-3}. And then cube roots, fourth roots, and general n^{th} roots. After this, you'll be ready for polynomials and solving polynomial equations.

Integer Exponents

EXPONENTIAL NOTATION AND TERMINOLOGY

Multiplication is repeated addition: $7 \cdot 4$ means $\underbrace{7 + 7 + 7 + 7}_{4 \text{ times}}$.

Similarly, **exponentiation**, or *raising* a number to a *power*, is repeated multiplication:

$$7^4 \text{ means } \underbrace{7 \cdot 7 \cdot 7 \cdot 7}_{4 \text{ times}}$$

In the expression 7^4, the number 7 is called the **base**; it's the number being multiplied. The 4 is the **exponent**; it tells how many times the base is multiplied by itself. We say that 7^4, or 2401, is the fourth **power** of 7.

This way of writing, with a base and a superscript exponent, is called **exponential notation**.

Exponential Notation

$$a^n = \underbrace{a \cdot a \cdot a \cdots a}_{n \text{ times}}$$

TRICKY POINTS

Powers of Negatives and Negatives of Powers Conventional order of operations indicates that exponentiation must be evaluated before addition or subtraction. In particular, this means that the expression -5^2 evaluates to -25: because the exponent is considered first, -5^2 means *the negative of the quantity five squared.*

To get *the square of the number negative five*, you have to use parentheses to put the minus sign first: $(-5)^2$ evaluates to 25.

Try these on your own before checking the solutions.

EXAMPLE: Exponential notation

Evaluate each of these expressions.

a. 6^3

b. -2.3^2

c. $-\left(\dfrac{1}{3}\right)^4$

d. $(-0.11)^3$

SOLUTION

a. $6^3 = 6 \cdot 6 \cdot 6 = 216$

b. $-2.3^2 = -(2.3)(2.3) = -5.29$
   Note that 2.3 is between 2 and 3, and 2.3^2 is between $2^2 = 4$ and $3^2 = 9$.

c. $-\left(\dfrac{1}{3}\right)^4 = \dfrac{1}{3} \cdot \dfrac{1}{3} \cdot \dfrac{1}{3} \cdot \dfrac{1}{3} = -\dfrac{1}{81}$

d. $(-0.11)^3 = (-0.11)(-0.11)(-0.11) = -0.001331$

SQUARES AND CUBES

The second power of a number is called the **square** of the number. The square of 4 is $4^2 = 16$. We can also say that 2.5 **squared** is $2.5^2 = 6.25$.

The third power of a number is called its **cube**. For example, the cube of 2 is $2^3 = 8$, and 10 **cubed** is $10^3 = 1000$.

The square of any real number (except 0) is positive. That's because the product of two positive numbers is positive and the product of two negative numbers is positive—so the product of any number with itself is positive. Any positive number is the square of two different numbers, one positive, the other negative. For example, 64 is the square of 8 and of –8, because $8^2 = 64$ and $(-8)^2 = 64$.

Cubes may be positive or negative. The cube of a positive number is positive, and the cube of a negative number is negative. For example, the cube of 3 is 27, but the cube of –3 is –27.

Perfect Squares, Perfect Cubes

The square of an integer is called a **perfect square**. (*Integers* are signed whole numbers: . . . , –2, –1, 0, 1, 2, . . .) For example, 81 is a perfect square because $81 = 9 \cdot 9$.

The cube of an integer is called a **perfect cube**. For example, 512 is a perfect cube because $512 = 8 \cdot 8 \cdot 8$.

Here are the first few perfect squares and cubes.

Squares	Cubes
$0^2 = 0$	$0^3 = 0$
$1^2 = 1$	$1^3 = 1$
$2^2 = 4$	$2^3 = 8$
$3^2 = 9$	$3^3 = 27$
$4^2 = 16$	$4^3 = 64$
$5^2 = 25$	$5^3 = 125$
$6^2 = 36$	$6^3 = 216$
$7^2 = 49$	$7^3 = 343$
$8^2 = 64$	$8^3 = 512$
$9^2 = 81$	$9^3 = 729$
$10^2 = 100$	$10^3 = 1000$

You can also talk about *perfect fourth powers, perfect fifth powers,* and so on. For example, 64 is a perfect sixth power: it is equal to 2^6.

EXPONENT RULES

There are several rules that make working with exponents and powers easier.

1. **Product of powers:** To multiply two powers of the same base, add their exponents.

> *Product of Powers*
>
> If *a* is any real number and *m* and *n* are exponents, then
>
> $$a^m a^n = a^{m+n}$$

That's because

$$a^m a^n = \left(\underbrace{a \cdot a \cdot a \cdots a}_{m \text{ times}} \right) \left(\underbrace{a \cdot a \cdot a \cdots a}_{n \text{ times}} \right) = \underbrace{a \cdot a \cdot a \cdots a}_{m+n \text{ times}}$$

For example,

$$(2^3)(2^5) = (2 \cdot 2 \cdot 2)(2 \cdot 2 \cdot 2 \cdot 2 \cdot 2)$$
$$= 2 \cdot 2 \cdot 2 \cdot 2 \cdot 2 \cdot 2 \cdot 2 \cdot 2$$
$$= 2^8$$

which is the same thing as 2^{3+5}.

2. **Quotient of powers:** To take the quotient of two powers of the same base, *subtract* their exponents.

> *Quotient of Powers*
>
> If *a* is any nonzero real number and *m* and *n* are exponents, then
>
> $$\frac{a^m}{a^n} = a^{m-n}$$

3. **Power of a power:** To raise a power to a power, multiply exponents:

> **Power of a Power**
> If a is a real number and m and n are exponents, then
> $$(a^m)^n = a^{mn}$$

That's because

$$(a^m)^n = \underbrace{(a^m)(a^m)\cdots(a^m)}_{n \text{ times}}$$

$$= \underbrace{\left(\underbrace{a \cdot a \cdots a}_{m \text{ times}}\right)\left(\underbrace{a \cdot a \cdots a}_{m \text{ times}}\right)\cdots\left(\underbrace{a \cdot a \cdots a}_{m \text{ times}}\right)}_{n \text{ times}}$$

$$= \underbrace{a \cdot a \cdot a \cdots a}_{mn \text{ times}}$$

For example,

$$(7^2)^3 = 7^2 \cdot 7^2 \cdot 7^2 = (7 \cdot 7)\,(7 \cdot 7)(7 \cdot 7) = 7^6 = 7^{2 \cdot 3}$$

4. **Power of a product:** The power of a product is the product of powers:

> **Power of a Product**
> If a and b are real numbers and n is an exponent, then
> $$(ab)^n = a^n b^n$$

That's because we can move factors around. For example,

$$\begin{aligned}
(2 \cdot 5)^3 &= (2 \cdot 5)(2 \cdot 5)(2 \cdot 5) \\
&= (2 \cdot 2 \cdot 2)(5 \cdot 5 \cdot 5) \\
&= 2^3 \cdot 5^3
\end{aligned}$$

5. **Power of a quotient:** Similarly, the power of a quotient is the quotient of powers:

> *Power of a Quotient*
> If a and b are real numbers, with
> $b \neq 0$, and n is an exponent, then
> $$\left(\frac{a}{b}\right)^n = \frac{a^n}{b^n}$$

6. **Exponent 1:** Any number to the first power is itself.

> *Exponent 1*
> If a is any real number, then
> $$a^1 = a$$

That's because a^1 is the product of 1 copy of a, which is just a.

7. **Exponent 0:** Any number to the zeroth power is 1. The only exception is 0 itself.

> *Exponent 0*
> If a is any nonzero real number, then
> $$a^0 = 1$$
> The expression 0^0 is undefined.

You can think of a^0 as the product of 0 copies of a. In a multiplicative world, we say that the product of nothing is 1: it has no interesting factors and doesn't change anything when you multiply by it.

TRICKY POINTS

0^0 *Is Undefined!* It may be tempting to say that 0^0 is 0 because 0 times anything is 0 again. And it may also be tempting to say that 0^0 is 1 because any number to the zeroth power is 1. But it's *because* there are two equally tempting answers that there can be no one right way to evaluate 0^0.

The expression 0^0 is undefined in the same way that dividing by 0 is undefined: evaluating either leads to havoc-wreaking contradictions.

EXAMPLE: Rules of exponents

Use properties of exponents in order to simplify each expression. You do not need to evaluate.

a. $3^7 \cdot 3^5$

b. $\dfrac{5^4 \cdot 5^7}{5^3}$

c. $(6^3)^5$

d. $\dfrac{(4^2 x)^3}{(4x)^2}$

e. $((17^8)^9)^0$

f. $\dfrac{4^{p+q}}{2^{2p}}$

SOLUTION

a. $3^7 \cdot 3^5 = 3^{7+5} = 3^{12}$

b. $\dfrac{5^4 \cdot 5^7}{5^3} = 5^{4+7-3} = 5^8$

c. $(6^3)^5 = 6^{3 \cdot 5} = 6^{15}$

d. $\dfrac{(4^2 x)^3}{(4x)^2} = \dfrac{(4^2)^3 x^3}{4^2 x^2} = 4^{2 \cdot 3 - 2} x^{3-2} = 4^4 x$

e. $((17^8)^9)^0 = 1$, because any nonzero number to the 0th power is 1

f. $\dfrac{4^{p+q}}{2^{2p}} = \dfrac{4^{p+q}}{(2^2)^p} = \dfrac{4^{p+q}}{4^p} = 4^{p+q-p} = 4^q$

NEGATIVE EXPONENTS

Making sense of a negative exponent like 3^{-1} or 5^{-2} can be tricky. But there's a way to think about it that's consistent with all the exponent rules. Take a look at these powers of 3:

Exponent	-3	-2	-1	0	1	2	3	4
Power	3^{-3}	3^{-2}	3^{-1}	3^0	3^1	3^2	3^3	3^4
	↓	↓	↓	↓	↓	↓	↓	↓
				1	3	9	27	81

Notice what's filled in in the bottom row. As you move right, these numbers increase by factors of 3. If you take 1 (which is 3^0) and multiply by 3, you get 3 (which is 3^1). If you take 3 (aka 3^1) and multiply by 3, you get 9 (which is 3^2). If you take 9 and multiply by 3, you get 27 (or 3^3). And if you take 27 and multiply by 3, you get 81 (which is 3^4).

Now reverse direction. As you move *left* in the table, the numbers in the bottom row *decrease* by factors of 3. Divide 81 by 3 and you get 27; divide 27 by 3 and you get 9; divide 9 by 3 and you get 3; divide 3 by 3 and you get 1.

So let's continue the pattern. Divide 1 by 3 and you get $\frac{1}{3}$. Let's write that in as 3^{-1}. Now divide $\frac{1}{3}$ by 3 and you get $\frac{1}{9}$; let's call that 3^{-2}. Divide $\frac{1}{9}$ by 3 and you get $\frac{1}{27}$; let's call that 3^{-3}.

Exponent	-3	-2	-1	0	1	2	3	4
Power	3^{-3}	3^{-2}	3^{-1}	3^0	3^1	3^2	3^3	3^4
	↓	↓	↓	↓	↓	↓	↓	↓
	$\frac{1}{27}$	$\frac{1}{9}$	$\frac{1}{3}$	1	3	9	27	81

You can continue this pattern for as long as you want: 3^{-4} should be $3^{-3} \div 3$, or $\frac{1}{81}$. And 3^{-5} is $3^{-4} \div 3$, or $\frac{1}{243}$. What this means is that you can compute any negative-integer power of 3.

Of course, people don't actually compute 3^{-100} by finding 3^{-99} and dividing it by 3: that's much too time-consuming. But take a look: according to the table, 3^1 is 3 and 3^{-1} is $\frac{1}{3}$. Similarly, 3^2 is 9 and 3^{-2} is $\frac{1}{9}$. And 3^3 is 27 and 3^{-3} is $\frac{1}{27}$.

See the pattern? To get 3^{-n}, divide 1 by 3^n. Said another way, 3^n and 3^{-n} are *reciprocals*: their product is 1. (And this too makes sense with the exponent rules: $3^n 3^{-n} = 3^{n + (-n)} = 3^0 = 1$.)

We can write this observation in general terms, for any base and any exponent:

> **Negative Exponents**
>
> If a is any nonzero real number and n is any integer, then
>
> $$a^{-n} = \frac{1}{a^n}$$

Try out this formula in a straightforward case.

EXAMPLE:

Evaluate 2^{-10}.

SOLUTION

$$2^{-10} = \frac{1}{2^{10}} = \frac{1}{1024}$$

KEY POINTS

Negative Exponents

1. The negative-exponent formula works with all exponent rules.

    For example, $2^4 2^{-6} = 2^{4-6}$, or 2^{-2}.

2. The formula works with fractional bases.

    For example, $\left(\frac{3}{5}\right)^{-1} = \frac{5}{3}$, and $\left(\frac{1}{4}\right)^{-3} = 4^3 = 64$.

3. The formula works in both directions, for positive exponents as well as for negative ones. What's key is that flipping a fraction over is the same thing as negating the exponent. For example, $\frac{1}{7^{-2}} = 7^{-(-2)} = 7^2$.

4. A negative exponent *does not* mean that the expression will evaluate to a negative number. In fact, it usually won't.

    a. If the base is positive, the power will always be positive: for example, $5^{-1} = \frac{1}{5}$.

    b. If the base is negative, the power will be positive if the exponent is even and negative if the exponent is odd.

    In short, only odd powers of negative bases are negative, regardless of whether the exponent is negative or positive.

EXAMPLE: Negative exponents I

Evaluate each expression.

a. 5^{-4}

b. $(-8)^{-3}$

c. $\dfrac{3^{-4}}{3^{-7}}$

SOLUTION

a. $5^{-4} = \dfrac{1}{5^4} = \dfrac{1}{625}$

b. $(-8)^{-3} = \dfrac{1}{(-8)^3} = -\dfrac{1}{512}$

c. $\dfrac{3^{-4}}{3^{-7}} = 3^{-4-(-7)} = 3^3 = 27$

EXAMPLE: Negative exponents II

Evaluate each expression.

a. $\left(\dfrac{1}{4}\right)^{-4}$

b. $\dfrac{2^{-3}}{7^{-2}}$

c. $\left(-\dfrac{2}{3}\right)^{-2}$

SOLUTION

a. $\left(\dfrac{1}{4}\right)^{-4} = 4^4 = 256$

b. $\dfrac{2^{-3}}{7^{-2}} = \dfrac{7^2}{2^3} = \dfrac{49}{8}$

c. $\left(-\dfrac{2}{3}\right)^{-2} = \left(-\dfrac{3}{2}\right)^2 = \dfrac{9}{4}$

Roots and Radicals

SQUARE ROOTS

In the world of positive numbers, the **square root** of a number a is the number whose square is a. For example, 7 is the square root of 49 because $7^2 = 49$.

The square root is denoted with a $\sqrt{}$-shaped **radical sign**, sometimes also called *square root sign*:

$$\sqrt{49} = 7$$

The whole expression $\sqrt{49}$ is called a **radical**.

The **radicand** is the number or expression under the radical sign. So the radicand in the expression $\sqrt{49}$ is 49; the radicand in the expression $3 + \sqrt{x^2 - 4}$ is $x^2 - 4$.

Finding the square roots of a number is also called *extracting* or *taking* its square roots.

EXAMPLE:

Find each square root.

a. $\sqrt{36}$

b. $\sqrt{1}$

c. $\sqrt{100}$

d. $\sqrt{169}$

e. $\sqrt{4900}$

SOLUTION

a. $\sqrt{36} = 6$

b. $\sqrt{1} = 1$

c. $\sqrt{100} = 10$

d. $\sqrt{169} = 13$

e. $\sqrt{4900} = 70$. You might be able to guess this one because you know that $7^2 = 49$, which means that

$$(7 \cdot 10)2 = 72 \cdot 102 = 49 \cdot 100$$

Square Roots and Negative Numbers

Now, let's discuss the square roots of negative numbers and negative square roots.

1. **Square roots of negative numbers:** As mentioned earlier in the chapter (page 148), the square of any real number is either positive or zero. What this means is that we can only take the square roots of nonnegative numbers. Negative numbers don't have square roots, as long as we're talking about real numbers. For example, the expression $\sqrt{-9}$ is undefined.

2. **Negative square roots:** We've said that $\sqrt{49}$ is 7 because $7^2 = 49$. But $(-7)^2$ is also 49, so what's wrong with saying that $\sqrt{49}$ is –7?

   Technically, nothing. Both 7 and –7 are square roots of 49. But we do want the expression $\sqrt{49}$ to represent a specific real number, not two real numbers; otherwise we wouldn't know how to interpret and evaluate expressions such as $3 + \sqrt{49}$. So we let $\sqrt{49}$ mean the positive square root of 49 *by definition*.

Every positive real number has exactly two square roots, one positive and one negative. But the radical sign, by convention, refers only to the positive square root, sometimes called the **principal square root**.

To refer to the negative square root, use a negative sign in front of the radical: $-\sqrt{49} = -7$. To refer to both square roots, you can either name them separately or use a $\pm$ sign. The two square roots of 49 are ± 7.

KEY POINTS

$\sqrt{a}$ *and the Square Roots of a*

- If a is positive, then $\sqrt{a}$ is the unique positive number whose square is a. The number a has two distinct square roots, $\sqrt{a}$ and $-\sqrt{a}$.
- If a is zero, there is only one square root: $\sqrt{0} = 0$.
- If a is negative, then $\sqrt{a}$ is undefined.
- The number a has no real number square roots.

EXAMPLE: Square roots and signs

Evaluate each of the following:

a. $\sqrt{64}$

b. $-\sqrt{64}$

c. $|\sqrt{64}|$

d. $|-\sqrt{64}|$

e. $-|\sqrt{64}|$

f. $\sqrt{|-64|}$

g. $\sqrt{-64}$

SOLUTION

a. $\sqrt{64} = 8$

b. $-\sqrt{64} = -8$

c. $|\sqrt{64}| = |8| = 8$

d. $|-\sqrt{64}| = |-8| = 8$

e. $-|\sqrt{64}| = -|8| = -8$

f. $\sqrt{|-64|} = \sqrt{64} = 8$

g. $\sqrt{-64}$ is undefined

Squares of Square Roots; Square Roots of Squares

Each positive number has two distinct square roots, so you can't always take the square root of the square of a number and expect to get the same number back. However, it's okay to go the other way. Here are both directions, with more details:

- **Square of a square root:** The square of a square root of a number is the number back again. For example, $\sqrt{36} = 6$ and $6^2 = 36$.

> *Square of a Square Root*
> If a is a nonnegative real number, then
> $$(\sqrt{a})^2 = a$$

- **Square root of a square:** The (principal) square root of a number is not necessarily the number you started with.

    - If the number is positive or zero, you will get the original number back. For example, $6^2 = 36$ and $\sqrt{36} = 6$.

    - If the number is negative, you will get the additive inverse of the number you started with. For example, $(-6)^2 = 36$, but $\sqrt{36} = 6$, not -6.

The square root of a square will always be nonnegative, regardless of the sign of the number that you start with. We can use absolute value to express this observation.

> *Square Root of a Square*
> If a is any real number, then
> $$\sqrt{a^2} = |a|$$

EXAMPLE: Square Roots of Squares

a. $\sqrt{(-8)^2}$

b. $\sqrt{-8^2}$

c. $-\sqrt{8^2}$

SOLUTION

a. $\sqrt{(-8)^2} = \sqrt{64} = 8$. Because the entire expression $(-8)^2$ is under the square root, you have to square -8 first, then extract the square root.

b. $\sqrt{-8^2} = \sqrt{-64}$, which is undefined. By order of operations, you have to square 8 first, then negate it, and only then try to take the square root.

c. $-\sqrt{8^2} = -\sqrt{64} = -8$. First square, then take the square root, then negate.

Estimating Square Roots

Every positive real number has a square root, but the vast majority of square roots are *irrational*—that is, they cannot be written down explicitly in decimal or fractional notation.

For example, $\sqrt{2}$ is a real number that may be written as a decimal, as

$$\sqrt{2} = 1.4142135623730950\ldots$$

but its decimal expansion continues on and on infinitely and never repeats. In practice, whenever we need to use $\sqrt{2}$ in a real-world computation, we approximate it with 1.4, or with 1.4142, or with 1.414213562, depending on the particular situation.

Although it's difficult to write down the precise digits of most square roots without using a calculator, we can still estimate that $\sqrt{2}$ is a number between $\sqrt{1}$ and $\sqrt{4}$, that is, between 1 and 2. Indeed, $\sqrt{2}$ is about 1.4, between 1 and 2.

Similarly, $\sqrt{30}$ is a number between $\sqrt{25}$ and $\sqrt{36}$, that is, between 5 and 6.

More formally, we can say that since

$$25 < 30 < 36$$

then

$$\sqrt{25} < \sqrt{30} < \sqrt{36}$$

which is equivalent to

$$5 < \sqrt{30} < 6$$

In this way, you can estimate the square root of any positive number by sandwiching the number between two consecutive perfect squares.

You can even go further and ask how $\sqrt{30}$ compares to 5.5. Since $5.5^2 = 30.25$, we know that $\sqrt{30}$ is quite close to but less than 5.5:

$$25 < 30 < 30.25$$

so

$$\sqrt{25} < \sqrt{30} < \sqrt{30.25}$$

or

$$5 < \sqrt{30} < 5.5$$

EXAMPLE: Estimate square roots

Estimate these square roots by sandwiching them between two consecutive perfect squares.

a. $\sqrt{57}$

b. $\sqrt{70}$

c. $\sqrt{8.5}$

d. $\sqrt{43.56}$

SOLUTION

a. Since $49 < 57 < 64$, then $\sqrt{49} < \sqrt{57} < \sqrt{64}$, which means that $7 < \sqrt{57} < 8$.

b. Since $64 < 70 < 81$, then $\sqrt{64} < \sqrt{70} < \sqrt{81}$, which means that $8 < \sqrt{70} < 9$.

c. Since $4 < 8.5 < 9$, then $\sqrt{4} < \sqrt{8.5} < \sqrt{9}$, which means that $2 < \sqrt{8.5} < 3$.

d. Since $36 < 43.56 < 49$, then $\sqrt{36} < \sqrt{43.56} < \sqrt{49}$, which means that $6 < \sqrt{43.56} < 49$. In fact, $\sqrt{43.56} = 6.6$, which you can check by computing 6.6^2.

CUBE ROOTS

The **cube root** of a real number a is the number whose cube is a. For example, 2 is the cube root of 8 because $2^3 = 8$. Also, –5 is the cube root of –125 because $(-5)^3 = -125$.

To denote cube roots, we use a square root sign and add a little 3 inside the hook, like so:

$$\sqrt[3]{8} = 2$$

We use the same terminology as with square roots. In the *radical* $\sqrt[3]{8}$, the *radical sign* $\sqrt[3]{}$ encloses the *radicand*.

Cube roots are more straightforward than square roots because there's no sign issue. Each real number has exactly one real-number cube root. Positive numbers have positive cube roots, negative numbers have negative cube roots, and the cube root of zero is zero.

EXAMPLE: Cube roots

Find each of these cube roots.

a. $\sqrt[3]{27}$

b. $\sqrt[3]{-1}$

c. $\sqrt[3]{1000}$

d. $\sqrt[3]{343}$

e. $\sqrt[3]{-8000}$

SOLUTION

a. $\sqrt[3]{27} = 3$

b. $\sqrt[3]{-1} = -1$

c. $\sqrt[3]{1000} = 10$

d. $\sqrt[3]{343} = 7$

e. $\sqrt[3]{-8000} = -20$

EXAMPLE: Estimate cube roots

Estimate each of these cube roots by sandwiching the radicand between two perfect cubes.

a. $\sqrt[3]{10}$

b. $\sqrt[3]{100}$

c. $\sqrt[3]{-5}$

SOLUTION

a. Since $8 < 10 < 27$, we have that $\sqrt[3]{8} < \sqrt[3]{10} < \sqrt[3]{27}$, or $2 < \sqrt[3]{10} < 3$. So $\sqrt[3]{10}$ is a number between 2 and 3.

b. Since $64 < 100 < 125$, we have that $\sqrt[3]{64} < \sqrt[3]{100} < \sqrt[3]{125}$, or $4 < \sqrt[3]{100} < 5$. So $\sqrt[3]{100}$ is somewhere between 4 and 5.

c. Since $-8 < -5 < -1$, we have that $\sqrt[3]{-8} < \sqrt[3]{-5} < \sqrt[3]{-1}$, or $-2 < \sqrt[3]{-5} < -1$. So $\sqrt[3]{-5}$ is a number between -1 and -2.

FOURTH ROOTS

A **fourth root** of a real number a is a number whose fourth power is a. For example, 2 is a fourth root of 16 because $2^4 = 16$.

We denote fourth roots just like cube roots, except with a little 4 instead of a little 3:

$$\sqrt[4]{16} = 2$$

Like square roots, fourth roots have problems with signs. The fourth power of any real number is positive or zero, so only non-negative numbers have fourth roots. Fourth roots of negative numbers are undefined.

By the same token, every positive number has two real-number fourth roots, one positive and one negative. For example, 2 and -2 are both fourth roots of 16. The radical $\sqrt[4]{16}$ isolates the positive *principal fourth root*.

EXAMPLE: Fourth roots

Find each of these fourth roots.

a. $\sqrt[4]{1}$

b. $\sqrt[4]{-1}$

c. $\sqrt[4]{81}$

d. $\sqrt[4]{10{,}000}$

SOLUTION

a. $\sqrt[4]{1} = 1$

b. $\sqrt[4]{-1}$ is undefined

c. $\sqrt[4]{81} = 3$

d. $\sqrt[4]{10{,}000} = 10$

GENERAL n^{th} ROOTS

More generally, an **n^{th} root** of a number a is a number whose n^{th} power is a. For example, 2 is a ninth root of 512 because $2^9 = 512$. And -3 is a sixth root of 729 because $(-3)^6 = 729$.

The radical sign $\sqrt[n]{\ }$ always denotes the so-called *principal n^{th} root*:

$$\sqrt[9]{512} = 2, \quad \sqrt[5]{-243} = -3, \quad \sqrt[6]{729} = 3$$

How do you know which n^{th} root is principal? It depends on whether n is odd or even. If you understand square roots and cube roots well, you can use them as models.

1. **Odd n:** If n is odd, each real number has exactly one real n^{th} root, so that's the principal root. Negative numbers have negative n^{th} roots and positive numbers have positive n^{th} roots.

   For example, 512 is positive, so its only real ninth root is positive; -243 is negative, so its only real fifth root is negative. Accordingly, we write $\sqrt[9]{512} = 2$ and $\sqrt[5]{-243} = -3$.

- If n is odd, the expression $\sqrt[n]{a}$ is defined for all real numbers a.

- Cube roots are n^{th} roots with odd n ($n = 3$). Like cube roots, n^{th} roots with odd n are straightforward about their signs.

2. **Even n:** If n is even, only nonnegative numbers have n^{th} roots. Each positive number has two n^{th} roots, one positive and the other negative. The positive one is the principal root.

   For example, both 3 and –3 are sixth roots of 729. But 3 is considered the principal root, so we write $\sqrt[6]{729} = 3$.

   - If n is even, the expression $\sqrt[n]{a}$ only makes sense for nonnegative a. Each positive a has two n^{th} roots, $\sqrt[n]{a}$ and $-\sqrt[n]{a}$.

   - Square roots are n^{th} roots with even n. (A square root has $n = 2$; you can think of $\sqrt{a}$ as shorthand for $\sqrt[2]{a}$). Like square roots, n^{th} roots with even n need delicate handling when it comes to plus and minus signs.

The number n is sometimes called the **index** of the nth root $\sqrt[n]{a}$. So a square root is a root of index 2 and a cube root is a root of index 3.

EXAMPLE: n^{th} roots

Evaluate each root expression.

a. $\sqrt[4]{2401}$

b. $\sqrt[11]{-2048}$

c. $\sqrt[3]{-1728}$

d. $\sqrt[27]{1}$

e. $\sqrt[58]{-1}$

SOLUTION

a. $\sqrt[4]{2401} = 7$

b. $\sqrt[11]{-2048} = -2$

c. $\sqrt[3]{-1728} = -12$

d. $\sqrt[27]{1} = 1$ because $1^{27} = 1$

e. $\sqrt[58]{-1}$ is undefined because 58 is even and only positive numbers have even-order roots.

Working with Radicals

MULTIPLYING AND DIVIDING SQUARE ROOTS

Multiplying Square Roots

What happens when you multiply two square roots together?

EXAMPLE: First square root multiply

Find $\sqrt{25} \cdot \sqrt{4}$.

SOLUTION

$$\sqrt{25} \cdot \sqrt{4} = 5 \cdot 2$$
$$= 10$$

Notice also that $25 \cdot 4 = 100$, and $\sqrt{100}$ just happens to be 10.

This example suggests a formula for multiplying square roots:

> *Products of Square Roots*
>
> If a and b are two nonnegative real numbers, then
>
> $$\sqrt{a}\sqrt{b} = \sqrt{ab}$$
>
> In other words, the product of square roots is the square root of the product.

EXAMPLE: Multiply square roots

Express each product as the square root of a single radicand.

a. $\sqrt{9}\sqrt{4}$

b. $\sqrt{1.44}\sqrt{100}$

c. $\sqrt{\left(\frac{1}{2}\right)}\sqrt{\left(\frac{2}{3}\right)}$

d. $\sqrt{3}\sqrt{x-7}$

e. $\sqrt{y+2}\sqrt{5y-1}$

f. $2\sqrt{3}$

SOLUTION

a. $\sqrt{9}\sqrt{4} = \sqrt{9\cdot 4} = \sqrt{36}$

b. $\sqrt{1.44}\sqrt{100} = \sqrt{1.44\cdot 100} = \sqrt{144}$

c. $\sqrt{\left(\frac{1}{2}\right)}\sqrt{\left(\frac{2}{3}\right)} = \sqrt{\left(\frac{1}{2}\right)\left(\frac{2}{3}\right)} = \sqrt{\frac{1}{3}}$

d. $\sqrt{3}\sqrt{x-7} = \sqrt{3(x-7)} = \sqrt{3x-21}$

e. $\sqrt{y+2}\sqrt{5y-1} = \sqrt{(y+2)(5y-1)} = \sqrt{5y^2+9y-2}$

f. $2\sqrt{3} = \sqrt{4}\sqrt{3} = \sqrt{4\cdot 3} = \sqrt{12}$

Dividing Square Roots

The rule for quotients of square roots is similar to the rule for products.

> #### Quotients of Square Roots
>
> If a is a nonnegative real number and b is a positive real number, then
>
> $$\frac{\sqrt{a}}{\sqrt{b}} = \sqrt{\frac{a}{b}}$$
>
> In other words, the quotient of two square roots is the square root of the quotient.

EXAMPLE: Divide square roots

Express each quotient as the square root of a single radicand.

a. $\dfrac{\sqrt{175}}{\sqrt{35}}$

b. $\dfrac{\sqrt{50x^2}}{\sqrt{5x}}$

c. $\dfrac{\sqrt{a^2b^3cd^4}}{\sqrt{abc^2}}$

d. $\dfrac{10}{\sqrt{5}}$

SOLUTION

a. $\dfrac{\sqrt{175}}{\sqrt{35}} = \sqrt{\dfrac{175}{35}} = \sqrt{5}$

b. $\dfrac{\sqrt{50x^2}}{\sqrt{5x}} = \sqrt{\dfrac{50x^2}{5x}} = \sqrt{10x}$

c. $\dfrac{\sqrt{a^2b^3cd^4}}{\sqrt{abc^2}} = \sqrt{\dfrac{a^2b^3cd^4}{abc^2}} = \sqrt{\dfrac{\overset{a}{\cancel{a^2}}\,\overset{b^2}{\cancel{b^3}}\,\cancel{c}d^4}{\cancel{ab}\cancel{c^2}_{c}}} = \sqrt{\dfrac{ab^2d^4}{c}}$

d. $\dfrac{10}{\sqrt{5}} = \dfrac{\sqrt{100}}{\sqrt{5}} = \sqrt{\dfrac{100}{5}} = \sqrt{20}$

KEY POINTS

Multiplying and Dividing Square Roots The multiplication and the division rules for square roots are very similar. That's because dividing is the same thing as multiplying by the reciprocal: for example, dividing a number by 3 and multiplying it by $\frac{1}{3}$ will give you the same result.

The reciprocal of a square root is the same thing as the square root of a reciprocal. For example:

$$\frac{1}{\sqrt{b}} = \sqrt{\frac{1}{b}}$$

The division rule is a special case of the multiplication rule.

EXAMPLE:
Use the multiplication rule to express $\frac{\sqrt{18}}{\sqrt{12}}$ as a single square root.

SOLUTION

$\frac{\sqrt{18}}{\sqrt{12}} = \sqrt{18} \cdot \frac{1}{\sqrt{12}}$ Dividing is multiplying by reciprocal

$\quad = \sqrt{18} \cdot \sqrt{\frac{1}{12}}$ Reciprocal of root is root of reciprocal

$\quad = \sqrt{18 \cdot \frac{1}{12}}$ Multiply square roots

$\quad = \sqrt{\frac{3}{2}}$ Simplify

MULTIPLYING AND DIVIDING n^{th} ROOTS

Multiplying and dividing higher-order roots works just like multiplying and dividing square roots. The product of two n^{th} roots is the nth root of the product, and the quotient of two n^{th} roots is the nth root of the quotient.

Products of Roots

If a and b are nonnegative real numbers and n is a natural number,

$$\sqrt[n]{a} \cdot \sqrt[n]{b} = \sqrt[n]{ab}$$

Quotients of Roots

If $b \neq 0$, then

$$\frac{\sqrt[n]{a}}{\sqrt[n]{b}} = \sqrt[n]{\frac{a}{b}}$$

Powers of Roots

Also,

$$\left(\sqrt[n]{a}\right)^m = \sqrt[n]{a^m}$$

Here and for the rest of the chapter, we're going to assume that anything that goes under a radical sign is positive or zero. This way we don't have to fret about whether n is even or odd, whether signs are positive or negative, and what kinds of values are allowed for variables.

EXAMPLE: Multiplying roots

Use the product rule to simplify each expression.

a. $\sqrt[3]{5} \cdot \sqrt[3]{7}$

b. $\sqrt[4]{4x} \cdot \sqrt[4]{10x^2}$

c. $\left(\sqrt[5]{3a}\right)^2$

SOLUTION

a. $\sqrt[3]{5} \cdot \sqrt[3]{7} = \sqrt[3]{5 \cdot 7} = \sqrt[3]{35}$

b. $\sqrt[4]{4x} \cdot \sqrt[4]{10x^2} = \sqrt[4]{4x \cdot 10x^2} = \sqrt[4]{40x^3}$

c. $\left(\sqrt[5]{3a}\right)^2 = \sqrt[5]{(3a)^2} = \sqrt[5]{9a^2}$

EXAMPLE: Dividing roots

Express each quotient as a single radical.

a. $\dfrac{\sqrt[3]{42}}{\sqrt[3]{14}}$

b. $\dfrac{\sqrt[7]{x^2 y^3}}{\sqrt[7]{x^{-2}}}$

c. $\dfrac{\sqrt[5]{2} \cdot \sqrt[5]{8}}{\sqrt[5]{2^2}}$

SOLUTION

a. $\dfrac{\sqrt[3]{42}}{\sqrt[3]{14}} = \sqrt[3]{\dfrac{42}{14}} = \sqrt[3]{3}$

b. $\dfrac{\sqrt[7]{x^2 y^3}}{\sqrt[7]{x^{-2}}} = \sqrt[7]{\dfrac{x^2 y^3}{x^{-2}}} = \sqrt[7]{x^{2-(-2)} y^3} = \sqrt[7]{x^4 y^3}$

c. $\dfrac{\sqrt[5]{2} \cdot \sqrt[5]{8}}{\sqrt[5]{2^2}} = \sqrt[5]{\dfrac{2 \cdot 8}{2^2}} = \sqrt[5]{4}$

SIMPLIFYING SQUARE ROOTS

You can use the multiplication rule for radicals to simplify radicals by pulling factors out of the radicand. Take a look at what you can do with $\sqrt{40}$:

$$\begin{aligned} \sqrt{40} &= \sqrt{2^2 \cdot 10} && \text{Factor } 40 = \text{(perfect square)} \cdot \text{(nonsquare)} \\ &= \sqrt{2^2} \cdot \sqrt{10} && \text{Separate the perfect square} \\ &= 2\sqrt{10} && \text{Simplify} \end{aligned}$$

The expression $2\sqrt{10}$ is considered more simplified than $\sqrt{40}$ because the radicand is a smaller number.

In general, a square-root radical is considered *simplified* if its radicand is not divisible by any perfect squares except 1. So $\sqrt{40}$ is not simplified because 40 is divisible by 4, which is a perfect square. But $2\sqrt{10}$ is simplified because 10 is not divisible by any perfect squares.

Another example, this time with variables: $\sqrt{2xy}$ is in simplified form because $2xy$ is not divisible by any perfect squares. On the other hand, $\sqrt{3xy^2}$ is not simplified because the radicand has a factor of y^2, which is a perfect square. Use the multiplication rule to simplify:

$$\sqrt{3xy^2} = \sqrt{3x}\sqrt{y^2}$$
$$= \sqrt{3x} \cdot y$$
$$= y\sqrt{3x}$$

So $\sqrt{3xy^2}$ can be rewritten as $y\sqrt{3x}$. Two brief notes about this example:

1. We usually write $y\sqrt{3x}$ rather than $\sqrt{3x}\,y$, to avoid confusing with $\sqrt{3xy}$.

2. The step $\sqrt{y^2} = y$ only works if y is positive or zero, not if it's negative. Here and elsewhere in this chapter we're assuming that the variables only take on nonnegative values if they're under a radical sign.

To simplify a square root:

1. Factor the radicand completely.

2. Group any factors that appear more than once in pairs: these form perfect squares.

3. Use the multiplication rule to separate the perfect squares from what's left over.

4. Simplify.

EXAMPLE: Simplify square root

Express $\sqrt{756}$ in simplified form.

SOLUTION

First, factor: $756 = 2^2 \cdot 3^3 \cdot 7$. Since 2 and 3 appear more than once, the radicand is divisible by perfect squares and the expression can be simplified.

Group factors in pairs: $756 = 2^2 \cdot 3^2 \cdot 3 \cdot 7$. Then use the multiplication rule for square roots to separate the paired factors from the unpaired:

$$\sqrt{756} = \sqrt{2^2 \cdot 3^2 \cdot 3 \cdot 7}$$
$$= \sqrt{2^2}\sqrt{3^2}\sqrt{3 \cdot 7}$$
$$= 2 \cdot 3 \cdot \sqrt{3 \cdot 7}$$
$$= 6\sqrt{21}$$

Since 21 is not divisible by any perfect squares, $6\sqrt{21}$ is in simplified form.

Simplifying works the same way when variables are involved.

EXAMPLE: Simplify square root with variables

Simplify $3\sqrt{800a^3bc^2}$.

SOLUTION

Factor and regroup the radicand:

$$800a^3bc^2 = 2^5 \cdot 5^2a^3bc^2 = 2^2 \cdot 2^2 \cdot 5^2 \cdot a^2 \cdot c^2 \cdot 2 \cdot ab$$

Then use the multiplication rule to find the simplified form:

$$3\sqrt{800a^3bc^2} = 3\sqrt{2^2 \cdot 2^2 \cdot 5^2 \cdot a^2 \cdot c^2 \cdot 2 \cdot ab}$$
$$= 3\sqrt{2^2}\sqrt{2^2}\sqrt{5^2}\sqrt{a^2}\sqrt{c^2}\sqrt{2ab}$$
$$= 3 \cdot 2 \cdot 2 \cdot 5 \cdot a \cdot c \cdot \sqrt{2ab}$$
$$= 60ac\sqrt{2ab}$$

SIMPLIFYING n^{th} ROOTS

An nth-root radical is considered **simplified** if its radicand is not divisible by any perfect n^{th} power. For example, $\sqrt[3]{4}$ and $\sqrt[4]{x^2y}$ are in simplified form, since 4 is not divisible by any cubes and x^2y is not divisible by any fourth powers. But $\sqrt[3]{16}$ and $\sqrt[4]{x^9y^3}$ are *not* simplified because 16 is divisible by 8, a perfect cube, and x^9y^3 is divisible by x^8, a perfect fourth power.

To simplify an nth root:

1. Factor the radicand and arrange any repeated factors in groups of n: these form perfect n^{th} powers.

2. Then use the multiplication rule to separate and simplify.

Take a look at how to simplify $\sqrt[3]{16}$ and $\sqrt[4]{x^9y^3}$:

$$\sqrt[3]{16} = \sqrt[3]{2^4} \qquad\qquad \text{Factor the radicand}$$

$$= \sqrt[3]{2^3 \cdot 2} \qquad\qquad \text{Group perfect cubes separately}$$

$$= \sqrt[3]{2^3} \cdot \sqrt[3]{2} \qquad\qquad \text{Separate}$$

$$= 2 \cdot \sqrt[3]{2} \qquad\qquad \text{Simplify}$$

So $\sqrt[3]{16}$ simplifies as $2\sqrt[3]{2}$.

$$\sqrt[4]{x^9y^3} = \sqrt[4]{x^8 \cdot xy^3} \qquad\qquad \text{Group factors}$$

$$= \sqrt[4]{x^8} \cdot \sqrt[4]{xy^3} \qquad\qquad \text{Separate perfect fourth power}$$

$$= \sqrt[4]{(x^2)^4} \cdot \sqrt[4]{xy^3} \qquad\qquad \text{Express as fourth power}$$

$$= x^2 \cdot \sqrt[4]{xy^3} \qquad\qquad \text{Simplify}$$

So $\sqrt[4]{x^9y^3} = x^2 \cdot \sqrt[4]{xy^3}$.

Try these on your own before reading the solutions.

EXAMPLE: Simplify n^{th} roots
Simplify each radical.

a. $4\sqrt{\dfrac{32x^5}{81}}$

b. $\sqrt[3]{-128y^5}$

SOLUTION

a. $4\sqrt{\dfrac{32x^5}{81}} = 4\sqrt{\dfrac{2^4x^42x}{3^4}}$ Find perfect fourth powers

$\qquad = 4\sqrt{\left(\dfrac{2x}{3}\right)^4 \cdot 2x}$ Combine perfect fourth powers

$\qquad = 4\sqrt{\left(\dfrac{2x}{3}\right)^4} \, \sqrt[4]{2x}$ Separate

$\qquad = \dfrac{2x}{3} \cdot \sqrt[4]{2x}$ Simplify

b. $\sqrt[3]{-128y^5} = \sqrt[3]{-4^3y^32y^2}$ Factor and find perfect cubes

$\qquad = \sqrt[3]{(-4y)^32y^2}$ Combine perfect cubes

$\qquad = \sqrt[3]{(-4y)^3} \cdot \sqrt[3]{(2y)^2}$ Separate

$\qquad = -4y \cdot \sqrt[3]{2y^2}$ Simplify

RATIONALIZING THE DENOMINATOR

Radicals in the denominator of a fractional expression are technically considered unsimplified. **Rationalizing the denominator** is a way to eliminate radicals from the denominator by moving them to the numerator.

Rationalizing Square Root Denominators

Let's see how rationalizing the denominator works with the fraction $\dfrac{2}{\sqrt{3}}$. To eliminate the $\sqrt{3}$ from the denominator, multiply both numerator and denominator by $\sqrt{3}$. (This is okay to do because multiplying both the numerator and the denominator of a fraction by the same thing doesn't change the value of the fraction. In this case, you're just multiplying by $\dfrac{\sqrt{3}}{\sqrt{3}}$, which is 1.)

$$\frac{2}{\sqrt{3}} = \frac{2 \cdot \sqrt{3}}{\sqrt{3}\sqrt{3}} \qquad \text{Multiply by } \frac{\sqrt{3}}{\sqrt{3}}$$

$$= \frac{2\sqrt{3}}{(\sqrt{3})^2} \qquad \text{Rewrite } \sqrt{3} \cdot \sqrt{3} = (\sqrt{3})^2$$

$$= \frac{2\sqrt{3}}{3} \qquad \text{Use } (\sqrt{3})^2 = 3$$

And $\frac{2\sqrt{3}}{3}$ is now in a form that is considered simplified.

Try these more complicated examples on your own before reading through the solutions. In each case, multiply the numerator and denominator by the square root in the denominator.

EXAMPLE: Rationalize denominator

Simplify $\frac{7\sqrt{3}}{3\sqrt{21}}$.

SOLUTION

$$\frac{7\sqrt{3}}{3\sqrt{21}} = \frac{7\sqrt{3} \cdot \sqrt{21}}{3\sqrt{21} \cdot \sqrt{21}}$$

$$= \frac{7\sqrt{3 \cdot 21}}{3 \cdot 21}$$

$$= \frac{7\sqrt{63}}{63}$$

The denominator has been rationalized, but there are repeated factors in the radicand, and the numerator and denominator have common factors. So the simplification can continue.

$$\frac{7\sqrt{63}}{63} = \frac{7\sqrt{3^2 \cdot 7}}{63}$$

$$= \frac{7 \cdot 3\sqrt{7}}{63}$$

$$= \frac{\cancel{7} \cdot \cancel{3}\sqrt{7}}{\underset{3}{\cancel{63}}}$$

$$= \frac{\sqrt{7}}{3}$$

OPTIONS

Canceling Square Root Factors In the previous example, there was a lot of multiplying and factoring out 3s and 7s. If you're feeling enterprising, you can save yourself some of this busywork.

For example, you may notice that $\dfrac{7\sqrt{3}}{3\sqrt{21}}$ has a factor of $\sqrt{3}$ in both numerator and denominator. So, it can be canceled out. Take a look—the canceled terms get crossed out:

$$\frac{7\sqrt{3}}{3\sqrt{21}} = \frac{7\cancel{\sqrt{3}}}{3\underset{\sqrt{7}}{\cancel{\sqrt{21}}}}$$

$$= \frac{7}{3\sqrt{7}}$$

Now you can rationalize the denominator:

$$\frac{7}{3\sqrt{7}} = \frac{7 \cdot \sqrt{7}}{3\sqrt{7} \cdot \sqrt{7}}$$

$$= \frac{7\sqrt{7}}{3\left(\sqrt{7}\right)^2}$$

$$= \frac{\cancel{7}\sqrt{7}}{3 \cdot \cancel{7}}$$

$$= \frac{\sqrt{7}}{3}$$

As you can see, the arithmetic is a little simpler. So long as every step you're doing is legitimate, you can do whatever you want in whatever order—you'll get the same answer eventually.

EXAMPLE: Rationalize with variables

Simplify $\dfrac{x^2\sqrt{x^3y^3}}{\sqrt{x^6y}}$.

SOLUTION

Notice that both the numerator and the denominator have

common factors of $\sqrt{x^3}$ and $\sqrt{y}$. They can be canceled out:

$$\frac{x^2\sqrt{x^3y^3}}{\sqrt{x^6y}} = \frac{x^2\cancel{\sqrt{x^3}}\cancel{\sqrt{y^3}}^{\sqrt{y^2}}}{\cancel{\sqrt{x^6}}_{\sqrt{x^3}}\cancel{\sqrt{y}}}$$

$$= \frac{x^2\sqrt{y^2}}{\sqrt{x^3}}$$

$$= \frac{x^2y}{\sqrt{x}}$$

$$= \frac{\cancel{x^2}^{x}\,y}{x\cancel{\sqrt{x}}}\cdot\frac{\sqrt{x}}{\sqrt{x}}$$

$$= \frac{xy\sqrt{x}}{x}$$

$$= y\sqrt{x}$$

Rationalizing n^{th} Root Denominators

What if you need to rationalize the denominator of $\dfrac{1}{\sqrt[3]{2}}$? If you

multiply both top and bottom by $\sqrt[3]{2}$, you'll get $\sqrt[3]{4}$ in the

denominator, which doesn't simplify to a rational number. But if

you multiply both top and bottom by $\sqrt[3]{4}$, things work out nicely:

$$\frac{1}{\sqrt[3]{2}} = \frac{1 \cdot \sqrt[3]{4}}{\sqrt[3]{2}\,\sqrt[3]{4}}$$
$$= \frac{\sqrt[3]{4}}{\sqrt[3]{8}}$$
$$= \frac{\sqrt[3]{4}}{2}$$

So $\dfrac{1}{\sqrt[3]{2}}$ is equal to $\dfrac{\sqrt[3]{4}}{2}$, whose denominator is rational.

In general, to eliminate a $\sqrt[n]{a}$ from the denominator, multiply

both numerator and denominator by $\sqrt[n]{a^{n-1}}$ or by a smaller n^{th}

root that will leave the denominator rational.

Try these examples on your own before reading the solutions.

EXAMPLE: Rationalizing n^{th} root denominator

Rationalize the denominator in each expression.

a. $\dfrac{y}{\sqrt[4]{y}}$

b. $\dfrac{2}{\sqrt[3]{9}}$

SOLUTION

a. Multiply both top and bottom by $\sqrt[4]{y^3}$ to rationalize.

$$\frac{y}{\sqrt[4]{y}} = \frac{y\sqrt[4]{y^3}}{\sqrt[4]{y}\,\sqrt[4]{y^3}}$$

$$= \frac{y\sqrt[4]{y^3}}{\sqrt[4]{y^4}}$$

$$= \frac{y\sqrt[4]{y^3}}{y}$$

$$= \frac{\cancel{y}\sqrt[4]{y^3}}{\cancel{y}}$$

$$= \sqrt[4]{y^3}$$

b. Multiplying both top and bottom by $\sqrt[3]{9^2}$ will definitely rationalize the denominator of $\dfrac{2}{\sqrt[3]{9}}$. But because $\sqrt[3]{9}$ is the same thing as $\sqrt[3]{3^2}$, you can get away with multiplying by $\dfrac{\sqrt[3]{3}}{\sqrt[3]{3}}$:

$$\frac{2}{\sqrt[3]{9}} = \frac{2\sqrt[3]{3}}{\sqrt[3]{9}\,\sqrt[3]{3}}$$

$$= \frac{2\sqrt[3]{3}}{\sqrt[3]{27}}$$

$$= \frac{2}{3}\sqrt[3]{3}$$

Summary

Exponents

1. Definition

- If n is positive, then

$$a^n = \underbrace{a \cdot a \cdot a \cdots a}_{n \text{ times}}$$

- If n is 0, for nonzero a,

$$a^0 = 1$$

- If n is negative, then $-n$ is positive and

$$a^n = \frac{1}{a^{-n}}$$

2. Properties

Product of powers	$a^m a^n = a^{m+n}$
Quotient of powers	$\dfrac{a^m}{a^n} = a^{m-n}$
Power of a power	$(a^m)^n = a^{mn}$
Power of a product	$(ab)^n = a^n b^n$
Power of a quotient	$\left(\dfrac{a}{b}\right)^n = \dfrac{a^n}{b^n}$

Radicals

1. Definitions

- **Square roots**

$$\sqrt{a} = b \qquad \text{if } b^2 = a,$$

and a and b are both nonnegative.

- **n^{th} roots**

$$\sqrt[n]{a} = b \qquad \text{if } b^n = a$$

and if n is even, a and b are both nonnegative.

2. Properties

Product of roots	$\sqrt{a}\sqrt{b} = \sqrt{ab}$	$\sqrt[n]{a} \cdot \sqrt[n]{b} = \sqrt[n]{ab}$
Quotient of roots	$\dfrac{\sqrt{a}}{\sqrt{b}} = \sqrt{\dfrac{a}{b}}$	$\dfrac{\sqrt[n]{a}}{\sqrt[n]{b}} = \sqrt[n]{\dfrac{a}{b}}$
Power of a root	$(\sqrt{a})^m = \sqrt{a^m}$	$\left(\sqrt[n]{a}\right)^m = \sqrt[n]{a^m}$

3. Simplifying radicals

- A square-root radical is *simplified* if the radicand is not divisible by any perfect square. Use $\sqrt{a^2} = a$ for positive a.

- An nth-root radical is simplified if the radicand is not divisible by any perfect n^{th} power. Use $\sqrt[n]{a^n} = a$ for positive a.

4. Rationalizing the denominator

- To get rid of $\sqrt{a}$ in the denominator, multiply the fraction by $\dfrac{\sqrt{a}}{\sqrt{a}}$.

- To get rid of $\sqrt[n]{a}$ in the denominator, multiply the fraction by $\dfrac{\sqrt[n]{a}}{\sqrt[n]{a}}$ or a simpler expression.

CHAPTER 4 EXPONENTS, ROOTS, RADICALS

Sample Test Questions

Answers to these questions begin on page 394.

1. Evaluate each expression.

   A. 5^{-2}

   B. $(-7)^{-3}$

   C. $\left(\dfrac{1}{8}\right)^{-2}$

   D. $\left(-\dfrac{3}{5}\right)^{-4}$

2. Simplify each expression.

   A. $\dfrac{n^2 m^{-3}}{n^4 m}$

   B. $\dfrac{(x^3)^{-4} x^5}{x^3 x^{-2}}$

   C. $\dfrac{\left(\dfrac{1}{2}\right)^{-3}}{\left(\dfrac{1}{4}y\right)^{-2}}$

   D. $\dfrac{3^m}{3^{-3}} \cdot \dfrac{9^{-m}}{3^{1-m}}$

3. Evaluate each expression.

   A. $\sqrt{-36}$

   B. $\sqrt{\dfrac{121}{49}}$

   C. $\sqrt{-(13)^2}$

   D. $\sqrt{|-13|^2}$

4. Evaluate each expression.

   A. $\sqrt[3]{-216}$

   B. $\sqrt[4]{-25^2}$

   C. $\sqrt[3]{\dfrac{27}{8}}$

   D. $\sqrt[9]{-8^3}$

5. Simplify each square root.

   A. $\sqrt{75}$

   B. $\sqrt{60}$

   C. $\sqrt{96xy^2}$

   D. $\sqrt{144x^3}$

6. Simplify each root.

   A. $\sqrt[3]{120}$

   B. $\sqrt[4]{x^7}$

   C. $\sqrt[3]{-64x^6}$

   D. $\sqrt[5]{256y^7}$

7. Simplify each expression.

   A. $\sqrt{6} \cdot \sqrt{10}$

   B. $\sqrt{18} \cdot \sqrt{32}$

   C. $\sqrt{15pq^2} \cdot \sqrt{5p^2}$

   D. $\dfrac{\sqrt{24xy^3}}{\sqrt{3x^3y^2}}$

8. Simplify each expression.

   A. $\sqrt{8} + \sqrt{2}$

   B. $\sqrt{2}(\sqrt{3} + \sqrt{5})$

   C. $\sqrt{12} + \sqrt{27} - \sqrt{3}$

   D. $\sqrt{6} \cdot \sqrt{15} - \sqrt{10}$

9. Rationalize the denominator of each expression.

   A. $\dfrac{4}{\sqrt{7}}$

   B. $\dfrac{-2\sqrt{3}}{\sqrt{5}}$

   C. $\dfrac{\sqrt{3} + 1}{\sqrt{2}}$

   D. $\dfrac{1 + \sqrt[3]{2}}{\sqrt[3]{2}}$

Polynomials and Factoring

5

Overview

Polynomials are special kinds of algebraic expressions. They're made out of sums and products of numbers and variables. For example, $3x^2 + 4x$ is a polynomial, as is $a^2 + 3a + 1$. They're useful to study for a couple of reasons. First, it turns out that many real-life situations can be described with polynomial equations. One important example is the trajectory of any object thrown into the air. Gravity pulls on the object in such a way that its height above the ground and its path are described by a polynomial equation.

Another reason why polynomials are so handy is that they're so much easier to work than many more complicated expressions. Scientists often use polynomials to approximate much nastier things. For example, the path of a swinging pendulum is not really described by a polynomial equation. But so long as the pendulum doesn't swing too far, a polynomial does a pretty good job of predicting where and when the pendulum will be.

Introducing Polynomials

A formal definition of the term *polynomial* can be confusing, so we'll start with a few examples:

$$3x^3 + 4x^2 - 6x - 10 \qquad 100z^{12} + z^2 - 4 \qquad 4x - \frac{7}{3}y + 3$$

$$7a^2b^4 - 6ab^3 - \frac{9}{2}a \qquad -3xyz^2 \qquad 45$$

Formally, a **monomial** is any product of powers of variables, possibly multiplied by a number. For example, $-3xyz^2$ and 45 are monomials. A **polynomial** is any monomial or sum or difference of monomials. A polynomial is an expression, not an equation; there should be no equal sign. All of the expressions above are polynomials.

A monomial that is part of a polynomial is called a **term**. For example, $3x^3 + 4x^2 - 6x - 10$ has four terms.

Terms that don't involve variables, like the –4 in $100z^{12} + z^2 - 4$, are called **constant terms**. Constant terms can be thought of as terms with a variable raised to the 0th power: for example, 45 is equivalent to $45x^0$ because $x^0 = 1$.

The number part of a term, including the sign, is called its **coefficient**. The coefficient of $-3xyz^2$ is –3. The coefficient of the third term of $3x^3 + 4x^2 - 6x - 10$ is –6.

Every polynomial is an algebraic expression, but not every algebraic expression is a polynomial. The algebraic expressions below are *not* polynomials.

$$\frac{1}{x} \qquad \sqrt{x} \qquad y^{\frac{1}{3}} \qquad y^{-4}$$

Polynomials cannot have terms with variables in the denominator of a fraction, variables under a radical sign, or variables raised to fractional or negative exponents. Only whole numbers can be exponents.

DEGREE OF A POLYNOMIAL

When we want to measure how big a number is, we take its absolute value. It turns out that the most useful way to measure how big a *polynomial* is is to take its *degree*.

The degree of a monomial term is the sum of the exponents of all the variables. So the degree of $4x^2$ is 2, the degree of $-3xyz^2$ is $1 + 1 + 2 = 4$, and the degree of $\frac{7}{2}y$ is 1. The degree of a constant term is 0—again, because you can think of that constant as multiplied by x^0.

The degree of a polynomial is the highest degree of any of its terms. So the degree of $7a^2b^4 - 6ab^3 - \frac{9}{2}a$ is 6, the degree of the first term. The degree of $4x - \frac{7}{3}y + 3$ is 1. Indeed, the degree of any linear expression is 1.

The exception is the polynomial 0, also called the *zero polynomial*. Although 0 is a constant, we say that the zero polynomial *has no degree*.

CLASSIFYING POLYNOMIALS BY DEGREE

There are special terms for polynomials of small degree.

Degree	Name for polynomial or term	Example
0	constant	-9
1	linear	$3x - 5y + 9$
2	quadratic	$4x^2 - 3x + 1$
3	cubic	$-x^3 - 7x^2 + 4x + 5$
4	quartic	$2x^4 - x^3 + x - 8$
5	quintic	$-x^5 + 3x^4 - 3x$

So one might say that $x^3 + 3x + 1$ is a *cubic polynomial*, that $z + 1$ is a *quartic*, or that the *linear term* of $xy^2 + 3y + 2$ has a coefficient of 3. One could even say that $x^2y + x + y$ is *quadratic in x*, meaning that the polynomial is quadratic if x is considered as the only variable.

The terms *quadratic*, *cubic*, *quartic*, and *quintic* are used almost exclusively for polynomials in one variable.

CLASSIFYING POLYNOMIALS BY NUMBER OF TERMS

You can also distinguish polynomials based on how many terms they have.

Degree	Name of polynomial	Example
one	monomial	$-2xy^3$
two	linear	$x^3 - 9$
three	quadratic	$3x^2 - xy + y^2$

Binomial is the most popular of these terms.

WRITING CONVENTIONS

Often, polynomials are organized in descending order of degree of terms, and each term is organized so that the variables go in alphabetical order:

$$4x^4 - 6x^2 + 8x + 9 \qquad ab^7 + 7a^2b^2 \qquad abc + ac + b$$

When this convention is used, if two terms have the same degree, then the one with the higher degree of the alphabetically earlier variable comes first:

$$4a^3b + 6a^2b^2 + 4ab^3 + 1$$

Polynomials in one variable are sometimes also organized in ascending order of degree, especially if the highest-degree coefficient is negative or if the variable is meant to assume very small values:

$$100 + 3a - 4a^2$$

POLYNOMIALS IN ONE VARIABLE

Polynomials in one variable are the easiest to understand.

A general n^{th}-degree polynomial in one variable looks like

$$ax^n + bx^{n-1} + cx^{n-2} + \cdots + dx^2 + ex + f$$

where a, b, c, d, e, f and any other coefficients are real numbers and $a \neq 0$.

For example, here's a one-variable polynomial of degree 9:

$$-2x^9 + 3x^8 + x^7 - x^5 + 4x^4 - 6x^3 - 11x^2 - x + 7$$

In a polynomial in one variable, the **leading term** is the highest-degree term, and the **lead coefficient** is the coefficient of the leading term. The leading term of the monster polynomial above is $-2x^9$; its lead coefficient is -2.

A **monic** polynomial has a lead coefficient of 1. For example, $x^2 - 4x + 4$ is a monic quadratic.

Polynomial Arithmetic

Polynomials can be added, subtracted, and multiplied. Polynomials in one variable may also be divided.

ADDING AND SUBTRACTING POLYNOMIALS

Any polynomial may be added to or subtracted from any other, but only so-called *like terms* may actually be combined.

Like Terms

Two terms are called *like terms* if they involve the same variables to the same powers. The number parts (that is, the *coefficients*) may be different. For example, a and $4a$ are like terms. So are $4xy^3$ and $-xy^3$. But $4xz^2$ and $4x^2z$ are *not* like terms because, for example, the first has a factor of x^2 and the second doesn't.

Like terms may not look alike if the variables are written in a different order. Arrange the variables in the same order before determining whether terms are like terms. Alphabetical order of variables is traditional, but any consistent system will do.

EXAMPLE: *Identify like terms*

Which of the following are like terms?

a. $-2x^2y^3$

b. $3y^3x^2$

c. $-4xyxyx$

d. $5yxyxy$

e. xy^2x^2

SOLUTION

We'll rearrange each term so that the xs come before the ys:

a. $-2x^2y^3$

b. $3y^3x^2 = 3x^2y^3$

c. $-4xyxyx = -4x^3y^2$

d. $5yxyxy = 5x^2y^3$

e. $xy^2x^2 = x^3y^2$

By comparing the number of xs and ys in each term, we see that (a) $-2x^2y^3$, (b) $3x^2y^3$, and (d) $5x^2y^3$ are like terms and that (c) $-4x^3y^2$ and (e) x^3y^2 are like terms.

Adding Polynomials

To add two polynomials, simply write their terms together. Then combine like terms by adding or subtracting their coefficients. The variables shouldn't change. For example, $-x^2y^3 + 5x^2y^3$ becomes $4x^2y^3$.

The distributive property is what makes this possible:

$$-x^2y^3 + 5x^2y^3 = (-1)x^2y^3 + (5)x^2y^3$$
$$= (-1 + 5)x^2y^3$$
$$= 4x^2y^3$$

Most people skip the intermediate steps or do them in their head.

EXAMPLE: Adding polynomials

Add $3x^2y - 5xy^2 + x^2 + xy + 3y$ and $8x^2y + 2xy^2 - xy + y^2 + 4$.

SOLUTION

$(3x^2y - 5xy^2 + x^2 + xy + 3y) + (8x^2y + 2xy^2 - xy + y^2 + 4)$

$= 3x^2y - 5xy^2 + x^2 + xy + 3y + 8x^2y + 2xy^2 - xy + y^2 + 4$

$= (3x^2y + 8x^2y) + (-5xy^2 + 2xy^2) + x^2 + (xy - xy) + y^2 + 3y + 4$

$= 11x^2y - 3xy^2 + x^2 + y^2 + 3y + 4$

Subtracting Polynomials

To subtract one polynomial from another, convert the subtraction to an addition by distributing the minus sign. In other words, flip the sign of every term in the polynomial being subtracted. Combine like terms to simplify the expression.

EXAMPLE: Subtracting polynomials

Subtract $-p^4 + 2p^2 + 6$ from $-p^4 + 3p^3 - 4p + 2$.

SOLUTION

$$(-p^4 + 3p^3 - 4p + 2) - (-p^4 + 2p^2 + 6)$$

$$= -p^4 + 3p^3 - 4p + 2 + p^4 - 2p^2 - 6$$

$$= (-p^4 + p^4) + 3p^3 - 2p^2 - 4p + (2 - 6)$$

$$= 3p^3 - 2p^2 - 4p - 4$$

MULTIPLYING POLYNOMIALS

Before multiplying polynomials, make sure that you know how to multiply expressions with exponents, like x^2y by x^3y^4.

> ### KEY POINTS
>
> *A Few Exponent Rules*
>
> 1. **Product of powers:** To multiply two powers with the same base, add their exponents:
>
> $$x^m \cdot x^n = x^{m+n}$$
>
> For example, $(x^3)(x^4) = x^{3+4} = x^7$. And $(x^2y)(x^2y^4) = x^4y^5$.
>
> 2. **Square of a power:** To square a power, multiply the exponent by 2:
>
> $$(x^n)^2 = x^{2n}$$
>
> For example, $(y^3)^2 = y^{3 \cdot 2} = y^6$.
>
> If you're squaring a product, you can square each factor separately:
>
> $$(x^n y^m)^2 = (x^n)^2 (y^m)^2 = x^{2n} y^{2m}$$
>
> For example, $(xy^3)^2 = (x)^2(y^3)^2 = x^2y^6$.
>
> See Chapter 4 for an in-depth review of working with exponents.

POLYNOMIALS AND FACTORING

CHAPTER 5

Two Monomials

To multiply two monomials, multiply the number parts and the variable parts separately. Use the rules of exponents to multiply powers of the same variable.

EXAMPLE: Product of monomials
Find the product of $3a^2bc^3$ and $-2a^4c^6d^2$.

SOLUTION

$$(3a^2bc^3)(-2a^4c^6d^2) = (3 \cdot (-2))(a^2 \cdot a^4)(b)(c^3 \cdot c^6)(d^2)$$
$$= -6a^6bc^9d^2$$

A Monomial and a Polynomial

To multiply a polynomial by a monomial, use the distributive property and multiply *each term* of the polynomial by the monomial. Simplify each term by combining powers of the same variable.

EXAMPLE: Product of polynomial and monomial
Multiply $10 - 3x - x^2$ by $2x^2$.

SOLUTION

$$(10 - 3x - x^2)2x^2 = (10)(2x^2) + (-3x)(2x^2) + (-x^2)(2x^2)$$
$$= 20x^2 - 6x^3 - 2x^4$$

OPTIONS

Spot-Check Your Work You can spot-check your work by choosing a value—any value—for the variable and making sure that the two sides agree when you plug in the value.

For example, if you want to check that $(10 - 3x - x^2)(2x^2)$ equals $20x^2 - 6x^3 - 2x^4$ in the example above, plug in, say, $x = 1$:

$$(10 - 3x - x^2)(2x^2) \overset{?}{=} 20x^2 - 6x^3 - 2x^4$$
$$(10 - 3(1) - (1)^2)(2(1)^2) \overset{?}{=} 20(1)^2 - 6(1)^3 - 2(1)^4$$
$$(6)(2) = 12$$

We get $(6)(2) = 12$, which is true, so the polynomial equation is likely to be correct as well.

EXAMPLE: Product of monomial and polynomial
Find the product of $-7y^2z$ and $-6y^2 + 2z^2 + yz - 9y + 3$.

SOLUTION
$$-7y^2z(-6y^2 + 2z^2 + yz - 9y + 3)$$
$$= (-7y^2z)(-6y^2) + (-7y^2z)(2z^2) + (-7y^2z)(yz)$$
$$+ (-7y^2z)(-9y) + (-7y^2z)(3)$$
$$= 42y^4z - 14y^2z^3 - 7y^3z^2 + 63y^3z - 21y^2z$$

Two Binomials

To multiply two binomials, you have to multiply each term of the first binomial by each term of the second binomial, for a total of four products. You can think through how this works by using the distributive property twice.

EXAMPLE: Product of binomials I
Find the product: $(x + 2)(y + 3)$.

SOLUTION
Treat $(y + 3)$ as a single unit and distribute it across the terms of $x + 2$:

$$(x + 2)(y + 3) = x(y + 3) + 2(y + 3)$$

Now you can use the distributive property on each of the resulting parenthesis pairs:

$$x(y + 3) + 2(y + 3) = xy + 3x + 2y + 6$$

So $(x + 2)(y + 3) = xy + 3x + 2y + 6$.

A general product of two binomials can be written as follows:

> *Product of Two Binomials*
> By the distributive property,
> $$(a + b)(c + d) = a(c + d) + b(c + d)$$
> $$= ac + ad + bc + bd$$
> Here, a, b, c, and d can be (signed) numbers or monomials.
>
> **FOIL:** common mnenomic
>
> **F**irst two (ac)
> **O**uter two (ad)
> **I**nner two (bc)
> **L**ast two (bd)

Although the distributed form of a product of two binomials has four terms, some of these terms may combine or cancel. A simplified product of two binomials will have two, three, or four terms. The degree of the product is the sum of the degrees of the two binomials.

EXAMPLE: Product of binomials II

Find the product of $x - 2y$ and $3x + y$.

SOLUTION

$$(x - 2y)(3x + y) = (x)(3x) + (x)(y) + (-2y)(3x) + (-2y)(y)$$
$$= 3x^2 + xy - 6xy - 2y^2$$
$$= 3x^2 - 5xy - 2y^2$$

EXAMPLE: Product of binomials III

Find the product: $(-pq^2 - 4p)(3q^2 - 5)$.

SOLUTION

$$(-pq^2 - 4p)(3q^2 - 5)$$
$$= (-pq^2)(3q^2) + (-pq^2)(-5) + (-4p)(3q^2) + (-4p)(-5)$$
$$= -3pq^4 + 5pq^2 - 12pq^2 + 20p$$
$$= -3pq^4 - 7pq^2 + 20p$$

Two Polynomials

The product of any two polynomials is computed in a similar way. Multiply every term of the first polynomial by every term of the second polynomial. Add all the product terms and simplify.

If one polynomial has n terms and the other m terms, then the product will initially have nm terms, though some of these may combine or cancel.

Use the distributive property to keep track of everything easily.

EXAMPLE: Product of polynomials

Simplify $(x^2 - 4x + 7)(2x^2 + 3x - 5)$.

SOLUTION

$(x^2 - 4x + 7)(2x^2 + 3x - 5)$

$= x^2(2x^2 + 3x - 5) + (-4x)(2x^2 + 3x - 5) + 7(2x^2 + 3x - 5)$

$= (x^2)(2x^2) + (x^2)(3x) + (x^2)(-5) + (-4x)(2x^2) + (-4x)(3x)$
$\quad + (-4x)(-5) + (7)(2x^2) + (7)(3x) + (7)(-5)$

(At this point, the product has 9 terms.)

$= 2x^4 + 3x^3 - 5x^2 - 8x^3 - 12x^2 + 20x + 14x^2 + 21x - 35$

$= 2x^4 + (3x^3 - 8x^3) + (-5x^2 - 12x^2 + 14x^2) + (20x + 21x) - 35$

$= 2x^4 - 5x^3 - 3x^2 + 41x - 35$

SPECIAL PRODUCTS

A few product types come up frequently, and it's nice to be able to recognize them.

Square of a Sum

EXAMPLE: Square of sum I

Find $(a + b)^2$.

SOLUTION

This square can be computed like any other product:

$(a + b)^2 = (a + b)(a + b) = (a)(a) + (a)(b) + (b)(a) + (b)(b)$
$\qquad\qquad\qquad\qquad = a^2 + ab + ab + b^2$
$\qquad\qquad\qquad\qquad = a^2 + 2ab + b^2$

It's good to memorize the statement $(a + b)^2 = a^2 + 2ab + b^2$.

> *Square of a Sum*
>
> If a and b are numbers or monomials, then
>
> $$(a + b)^2 = a^2 + 2ab + b^2$$
>
> In other words, the square of a sum of two terms is the square of the first term, plus twice the product of the terms, plus the square of the second term.

EXAMPLE: Square of sum II
Find $(2x + 3)^2$.

SOLUTION
Using the formula with $a = 2x$ and $b = 3$,

$$(2x + 3)(2x + 3) = (2x)^2 + 2(2x)(3) + (3)^2$$
$$= 4x^2 + 12x + 9$$

Square of a Difference

EXAMPLE: Square of difference
Find $(4x - 1)^2$.

SOLUTION
Using the formula with $a = 4x$ and $b = -1$,

$$(4x - 1)(4x - 1) = (4x)^2 + 2(4x)(-1) + (-1)^2$$
$$= 16x^2 - 8x + 1$$

In this example, b is negative. The formula $(a + b)^2 = a^2 + 2ab + b^2$ works for both positive and negative a and b, but you can also memorize a formula for the square of a difference.

> *Square of a Difference*
>
> $$(a - b)^2 = a^2 - 2ab + b^2$$
>
> Here, a and b can be numbers or monomials.

Only the so-called **cross-term**—the middle term that involves both as and bs—has changed sign. The b^2 isn't negative because $(-b) \cdot (-b) = b^2$.

Difference of Squares

The product of polynomials in the form $(a + b)(a - b)$ has a particularly nice form.

EXAMPLE: Difference of squares I
Find $(a + b)(a - b)$.

SOLUTION

$$\begin{aligned}(a + b)(a - b) &= (a)(a) + (a)(-b) + (b)(a) + (b)(-b) \\ &= a^2 - ab + ab - b^2 \\ &= a^2 - b^2\end{aligned}$$

The cross-terms cancel, leaving $(a + b)(a - b)$ equal to a **difference of squares**.

> **Difference of Squares**
>
> If a and b are numbers or monomials, then
>
> $$(a + b)(a - b) = a^2 - b^2$$

EXAMPLE: Difference of squares II
Simplify $(3x + 4y)(3x - 4y)$.

SOLUTION
Use the difference of squares formula with $a = 3x$ and $b = 4y$:

$$\begin{aligned}(3x + 4y)(3x - 4y) &= (3x)^2 - (4y)^2 \\ &= 9x^2 - 16y^2\end{aligned}$$

POLYNOMIAL LONG DIVISION

You can divide one polynomial in one variable by another using a process of long division similar to long division of large numbers.

But if you're not going to include the whole division problem, simply do:

$$8 \overline{\smash{)}\ 3014} \\ \ \ \ 376$$

To review, in the equation

$$3014 \div 8 = 376$$

3014 is the *dividend*, 8 is the *divisor*, and 376 is the *quotient*.

In polynomial long division, the terms of the polynomial play the same role that digits play when you're dividing integers.

The best way to learn polynomial long division is to work through an example.

EXAMPLE: Polynomial long division

Find the quotient and remainder when $x^3 - 5x^2 + 4x + 3$ is divided by $x - 2$.

SOLUTION

1. Arrange both polynomials in descending order of degree and set up the long division. If there are any zero coefficients, use a $0x^k$ placeholder term.

$$x - 2 \,\overline{\big)\, x^3 - 5x^2 + 4x + 3}$$

2. Divide the first term of the dividend (x^3) by the first term of the divisor (x). Write the result down as the first term of the quotient.

$$\begin{array}{r} x^2 \\ x - 2 \,\overline{\big)\, x^3 - 5x^2 + 4x + 3} \end{array}$$

3. Multiply the divisor by the quotient term. Write the product under the dividend, aligning degrees.

$$\begin{array}{r} x^2 \\ x - 2 \,\overline{\big)\, x^3 - 5x^2 + 4x + 3} \\ x^3 - 2x^2 \end{array}$$

4. Subtract the product from the dividend. The first terms should cancel, leaving a difference with a smaller degree than the dividend.

$$\begin{array}{r} x^2 \\ x - 2 \,\overline{\big)\, x^3 - 5x^2 + 4x + 3} \\ \underline{-(x^3 - 2x^2)} \\ -3x^2 + 4x + 3 \end{array}$$

5. Repeat, using the difference from the previous step as the new dividend. The division is done when you can't divide (while staying polynomial) anymore.

$$
\begin{array}{r}
x^2 - 3x - 2 \\
x - 2 \overline{) x^3 - 5x^2 + 4x + 3} \\
\underline{-(x^3 - 2x^2)} \\
-3x^2 + 4x + 3 \\
\underline{-(-3x^2 + 6x)} \\
-2x + 3 \\
\underline{-(-2x + 4)} \\
-1
\end{array}
$$

Since you can't divide -1 by x (and get a polynomial result), the division is over.

6. The last difference is the remainder. Its degree *must* be strictly less than the degree of the divisor.

So $x^3 - 5x^2 + 4x + 3$ divided by $x - 2$ is $x^2 - 3x - 2$, remainder -1.

Note that the degree of the remainder (degree of the constant polynomial -1 is 0) is strictly less than the degree of the divisor (degree of $x - 2$ is 1).

To check the division, multiply the divisor by the quotient and add the remainder: you should get the dividend. That is, you should check that

$$(x - 2)(x^2 - 3x - 2) + (-1) = x^3 - 5x^2 + 4x + 3$$

Try the next example on your own before looking at the solution.

EXAMPLE: Long division

Divide $6x^4 - x^3 + 9x + 2$ by $2x + 1$.

SOLUTION

The dividend has no x^2 term, so add the placeholder term $0x^2$.
Set up the long division.

$$
\begin{array}{r}
3x^2 - 2x^2 + \;\; x + 4 \\
2x + 1 \overline{) 6x^4 - \;\; x^3 + 0x^2 + 9x + 2} \\
\underline{-(6x^4 + 3x^3)} \\
-4x^3 + 0x^2 + 9x + 2 \\
\underline{-(-4x^3 - 2x^2)} \\
2x^2 + 9x + 2 \\
\underline{-(2x^2 + \;\; x)} \\
8x + 2 \\
\underline{-(8x + 4)} \\
-2
\end{array}
$$

The quotient is $3x^3 - 2x^2 + x + 4$, and the remainder is -2.
Said another way, we've computed that

$$
\frac{6x^4 - x^3 + 9x + 2}{2x + 1} = 3x^3 - 2x^2 + x + 4 + \frac{-2}{2x + 1}
$$

Such polynomial quotients are called *rational expressions*.

Synthetic Division **Synthetic division** is an orderly method of recording coefficients in polynomial long division; it works when the divisor is in the form $x - a$. The downside in this cleaner-looking method is that it's more difficult to see what's going on and easier to make a mistake.

The best way to learn synthetic division is to work through an example.

EXAMPLE:
Divide $x^4 - 5x^3 + 14x + 8$ by $x - 3$.

Synthetic division uses three lines of coefficients. The first records the divisor and the dividend, the second is a helper line, and the third records the quotient and the remainder.

When dividing a polynomial by $x - a$, write a in the left-hand corner of the top line; next to it write the coefficients of the dividend, in descending order of degree and padded with 0s if necessary. A polynomial of degree n should take $n + 1$ coefficients to write. Leave a space for the helper numbers and draw a horizontal line.

SOLUTION
Set up the first line of synthetic division by recording the coefficients of $x^4 - 5x^3 + 14x + 8$; write 3 in the left-hand corner because you're dividing by $x - 3$.

$$\underline{3\rfloor \quad\quad 1 \quad\quad -5 \quad\quad 0 \quad\quad 14 \quad\quad 8}$$

The 1 is shorthand for x^4, the -5 for $-5x^3$, the 0 for $0x^2$, the 14 for $14x^1$, and the 8 for $8x^0$.

Copy the first coefficient of the dividend under the line. Multiply it by a and write the product above the line under the second coefficient of the dividend.

$$\begin{array}{c|ccccc} 3\rfloor & 1 & -5 & 0 & 14 & 8 \\ & & 3 & & & \\ \hline & 1 & & & & \end{array}$$

Add the second coefficient and the number below and write the sum under the line. Then repeat the process. Multiply the new number under the line by a, and write the product under the next coefficient of the dividend.

$$\begin{array}{r|rrrrr} 3 & 1 & -5 & 0 & 14 & 8 \\ & & 3 & -6 & & \\ \hline & 1 & -2 & & & \end{array}$$

Repeat until you run out of coefficients. The remainder is the last number under the line; the other numbers under the line spell out the quotient. The degree of the quotient is one less than the degree of the dividend.

$$\begin{array}{r|rrrrr} 3 & 1 & -5 & 0 & 14 & 8 \\ & & 3 & -6 & -18 & -12 \\ \hline & 1 & -2 & -6 & -4 & -4 \end{array}$$

So the remainder here is –4 and the quotient is $x^3 - 2x^2 - 6x - 4$.

Factoring Polynomials

When you factor whole numbers, you take a bigger number and write it as a product of smaller numbers. For example, you can factor 12 as $3 \cdot 4$ or as $2 \cdot 2 \cdot 3$.

When you **factor** a polynomial, you're doing something similar: you take a bigger-degree polynomial and write it as a product of smaller-degree polynomials. For example, the polynomial $x^2 + 4x + 3$ can be factored as $(x + 1)(x + 3)$. The polynomial $x^4 - 2x^3 + 4x^2 - 6x + 3$ factors as $(x - 1)^2(x^2 + 3)$.

Factoring unfamiliar polynomials can be difficult. Not all polynomials factor, and there's no single method that tells you how to do it even when they do. Factoring is a bit of an art: you may have to try a little of this and a little of that. Here are some tricks of the trade.

PULLING OUT COMMON FACTORS

When faced with a polynomial to factor, the first thing to do is look for factors common to all the terms. Check the coefficients: find the **greatest common factor** of all the coefficients and factor it out. Check all the variables: if any variable appears in every term, factor it out. You should be able to pull out the smallest power that appears in any term.

The remaining polynomial should have two features: its coefficients should have no common factors (except for 1), and there should be no variable that appears in every term.

KEY POINTS

Greatest Common Factor The *greatest common factor* (GCF) of two or more integers is the greatest of the factors that all of them have in common. For example, 6 is a common factor of 24 and 84, but their greatest common factor is 12.

To find the GCF of two or more integers, first find their prime factorizations. Each prime that appears in *every* prime factorization must appear in the GCF, and primes that appear several times in every factorization must appear several times in the GCF.

For example, to find the GCF of 90 and 24, factor both: $90 = 2 \cdot 3^2 \cdot 5$ and $54 = 2 \cdot 3^3$. The two primes that appear in both factorizations are 2 and 3, so both will appear in the GCF. Since 3 appears at least twice in both 90 and 54, 3^2 must appear in the GCF. So, the GCF of 90 and 54 is $2 \cdot 3^2 = 18$.

EXAMPLE: Factor out common factors

Factor out the common factors from $36x^4 + 48x^3 - 60x^2$.

SOLUTION

The (unsigned) coefficients are $36 = 2^2 \cdot 3^2$, $48 = 2^4 \cdot 3$ and $60 = 2^2 \cdot 3 \cdot 5$. So the greatest common factor of all three of the coefficients is $2^2 \cdot 3 = 12$.

Every term has a factor of x; the greatest power of x that appears anywhere is x^2.

Factoring out $12x^2$ gives

$$36x^4 + 48x^3 - 60x^2 = (12x^2)(3x^2) + (12x^2)(4x) - (12x^2)(5)$$
$$= 12x^2(3x^2 + 4x - 5)$$

EXAMPLE: Factor out common factors and –1

Pull out common factors from $-4y^4z^2 + 27y^3z - 6y^2z^2$.

SOLUTION

The (unsigned) coefficients are $4 = 2^2$, $27 = 3^3$, and $6 = 2 \cdot 3$. No primes appear in all three, so the greatest common factor is 1.

Every term has a factor of y; the greatest power of y appearing everywhere is y^2.

Every term has a factor of z; the smallest power of z anywhere is z.

So we factor out y^2z:

$$-4y^4z^2 + 27y^3z - 6y^2z^2 = y^2z(-4y^2z) + y^2z(27y) + y^2z(-6z)$$
$$= y^2z(-4y^2 + 27y - 6z)$$

So $-4y^4z^2 + 27y^3z - 6y^2z^2 = y^2z(-4y^2 + 27y - 6z)$.

Though we're done with the problem, the main polynomial $-4y^2 + 27y - 6z$ still has a negative lead coefficient, which many people dislike. Let's factor out a –1 just for practice; every term should switch sign:

$$-4y^2 + 27y - 6z = -1(4y^2 - 27y + 6z)$$

So we can also write

$$-4y^4z^2 + 27y^3z - 6y^2z^2 = -y^2z(4y^2 - 27y + 6z)$$

FACTORING BY GROUPING

Factoring by grouping is a factoring technique that works best on four-term polynomials that happen to be products of two binomials. Take a look at how it works in this example.

EXAMPLE: Factoring by grouping
Factor $2ac - bc + 6ad - 3bd$.

SOLUTION
Make two groups of two. Start out naively, grouping the first two terms and the last two terms:

$$2ac - bc + 6ad - 3bd = (2ac - bc) + (6ad - 3bd)$$

Pull out common factors from each group. The first two terms have a common factor of c. The other two have a common factor of $3d$:

$$(2ac - bc) + (6ad - 3bd) = c(2a - b) + 3d(2a - b)$$

The leftovers are the same in both groups. Now you can undistribute once again, pulling out the common factor of $(2a - b)$ from each group:

$$c(2a - b) + 3d(2a - b) = (c + 3d)(2a - b)$$

So $2ac - bc + 6ad - 3bd = (c + 3d)(2a - b)$.

That's the general method of factoring by grouping: arrange the four terms in two groups of two and pull out common factors from each group. The hope is that the leftover pieces in each group will be the same or perhaps differ by a sign.

Factoring by grouping depends on the order of the terms, so sometimes you may have to rearrange things for the method to work. If the polynomial doesn't seem to factor nicely with the order of terms as given, switch the middle two terms and try again. If that also doesn't work, the polynomial cannot be factored by grouping.

EXAMPLE: Factoring by grouping with rearrangement

Factor $6xy + 5 - 10x - 3y$.

SOLUTION

First, try grouping the first term with the second and the third with the fourth:

$$6xy + 5 - 10x - 3y = (6xy + 5) + (-10x - 3y)$$

Neither the first group nor the second has any common factors. This grouping doesn't lead to a factorization. Let's swap the middle two terms and regroup:

$$6xy - 10x + 5 - 3y = (6xy - 10x) + (5 - 3y)$$

The first group has a common factor of $2x$. The second group doesn't have any common factors, but that's fine:

$$(6xy - 10x) + (5 - 3y) = 2x(3y - 5) + (5 - 3y)$$

Although the leftover parts are not identical, they are very close: they differ by a factor of -1. Factor -1 out of the second group to flip the signs of the terms:

$$2x(3y - 5) + (5 - 3y) = 2x(3y - 5) + (-1)(3y - 5)$$

Finally, finish undistributing by pulling out $(3y - 5)$:

$$2x(3y - 5) + (-1)(3y - 5) = (2x - 1)(3y - 5)$$

So $6xy + 5 - 10x - 3y = (2x - 1)(3y - 5)$.

EXAMPLE: Factoring by grouping, with common factors

Factor $-2x^4 + 4x^3 - 6x^2 + 12x$.

SOLUTION

First, pull out common factors. All the coefficients have a factor of 2; every term has a factor of x. Since the leading term is negative, we'll also pull out a negative sign.

Factoring out $-2x$ gives:

$$-2x^4 + 4x^3 - 6x^2 + 12x = -2x(x^3 - 2x^2 + 3x - 6)$$

Now we can try to factor $x^3 - 2x^2 + 3x - 6$ by grouping:

$$
\begin{aligned}
x^3 - 2x^2 + 3x - 6 &= (x^3 - 2x^2) + (3x - 6) \\
&= x^2(x - 2) + 3(x - 2) \\
&= (x^2 + 3)(x - 2)
\end{aligned}
$$

So $-2x^4 + 4x^3 - 6x^2 + 12x = -2x(x^2 + 3)(x - 2)$.

When Factoring by Grouping Fails

Because of the way signs multiply, a polynomial that can be factored by grouping *must* have an even number of negative terms. All four terms may be positive, two may be positive and two negative, or all four terms may be negative.

So if a polynomial has one or three negative terms, you can conclude that it cannot be factored by grouping. You don't even need to bother with rearrangements.

EXAMPLE:

Factor $xy + 2y - 3x + 6$.

SOLUTION

It's tempting to factor this polynomial by grouping—that y! That 3! That 6! But there's only one negative term, an odd number, so factoring by grouping won't work. Take a look:

$$xy + 2y - 3x + 6 = (xy + 2y) + (-3x + 6)$$
$$= y(x + 2) + 3(-x + 2)$$

And now there's nothing you can do. Even though $x + 2$ and $-x + 2$ look similar, they're not the same polynomial and they don't differ by a factor of -1. This polynomial simply doesn't factor. Counting the negative signs could have tipped you off and saved you some work.

FACTORING $x^2 + bx + c$

Factoring by grouping is a great method, but you don't always know what the four terms are because some of them combine together.

For example, the product $(x + p)(x + q)$ multiplies out to $x^2 + px + qx + pq$. In this form, factoring by grouping is straight-forward: you get the factorization $(x + p)(x + q)$ quickly. But unfortunately $x^2 + px + qx + pq$ usually appears in the form $x^2 + (p + q)x + pq$, with only three terms. To factor it, we have to recover p and q just by looking at $p + q$ and pq, the coefficients of $x^2 + (p + q)x + pq$.

Looking at it from the other direction, we can try to factor a polynomial in the form $x^2 + bx + c$ by looking for two (signed) numbers p and q whose (signed) sum is $p + q = b$ and whose (signed) product is $pq = c$. Once we know p and q, we know the factorization: $x^2 + bx + c = (x + p)(x + q)$.

Finding p and q is all about smart trial and error. Usually, you list all the pairs of numbers whose product is c and choose a pair whose sum is b.

EXAMPLE: Factoring $x^2 + bx + c$

Factor $x^2 + 9x + 20$.

SOLUTION

We're looking for p and q whose product is 20 and whose sum is 9.

List all the factorizations of 20: $20 = 1 \cdot 20$, $20 = 2 \cdot 10$, and $20 = 4 \cdot 5$. Now check each one:

- 1 and 20: sum is 21. No good.

- 2 and 10: sum is 12. No good.

- 4 and 5: sum is 9. Good.

So $x^2 + 9x + 20 = (x + 4)(x + 5)$.

Check by multiplying:

$$(x + 4)(x + 5) = x^2 + 5x + 4x + 20 = x^2 + 9x + 20.$$

Things are only a little more complicated when negative signs enter the picture.

EXAMPLE: Factoring $x^2 - bx - c$

Factor $x^2 - 2x - 15$.

SOLUTION

We're looking for two signed numbers p and q whose product is −15 and whose signed sum is −2. List the pairs of numbers whose product is −15 and check their sum.

- −1 and 15: sum is 14. Nope.

- −15 and 1: sum is −14. Nope.

- −3 and 5: sum is 2. Nope.

- −5 and 3: sum is −2. Great.

So p and q are −5 and 3, and $x^2 - 2x - 15 = (x - 5)(x + 3)$.
    Check: $(x - 5)(x + 3) = x^2 + 3x - 5x - 15 = x^2 - 2x - 15$.

CONSIDER THE SIGNS

We can bypass some of the guessing and checking by looking at the signs of the sum and product of p and q. In this case, the sum is −2 and the product is −15.

- Since their product is negative, p and q must have different signs. This means that their sum is the *difference* in their absolute values.

- Since their sum is negative, the one with the *greater* absolute value is negative and the other one is positive.

So we just look for the absolute values of p and q: two positive numbers whose product is 15 and whose difference is 2. That's 3 and 5. The greater one is negative, so the factorization is $(x - 5)(x + 3)$.

EXAMPLE: Factoring $x^2 - bx + c$

Factor $x^2 - 15x + 36$.

SOLUTION

We're looking for two numbers p and q whose (signed) product is 36 and whose (signed) sum is –15. List all the pairs of signed numbers whose product is 36 and check their sum.

- 1 and 36: sum is 37
- –1 and –36: sum is –37
- 2 and 18: sum is 20
- –2 and –18: sum is –20
- 3 and 12: sum is 15
- –3 and –12: sum is –15. That's it. We can stop now.
- 4 and 9
- –4 and –9
- 6 and 6
- –6 and –6

So $x^2 - 15x + 36 = (x - 3)(x - 12)$.

Check by multiplying: $(x - 3)(x - 12) = x^2 - 12x - 3x + 36 = x^2 - 15x + 36$.

CONSIDER THE SIGNS

Again, we can eliminate some of the guesswork by looking at the signs of the sum and product.

- The product is positive, so p and q have the same sign. This means that their sum is the *sum* of their absolute values.
- The sum is negative, so p and q are both negative.

Now find the absolute values of p and q: two positive numbers whose product is 36 and whose sum is 15. That's 3 and 12. Since both are negative, the factorization is $(x - 3)(x - 12)$.

EXAMPLE: Factoring $x^2 + bx - c$

Factor $x^2 + 7x - 30$.

SOLUTION

We're looking for two numbers whose product is –30 and whose sum is 7. Check the factorizations of –30:

- 1 and –30: sum is –29

- –1 and 30: sum is 29

- 2 and –15: sum is –13

- –2 and 15: sum is 13

- 3 and –10: sum is –7

- –3 and 10: sum is 7. That's it.

- 5 and –6

- –5 and 6

So –3 and 10 work, and $x^2 + 7x - 30$ factors as $(x + 10)(x - 3)$.
Check: $(x + 10)(x - 3) = x^2 - 3x + 10x - 30 = x^2 + 7x - 30$.

CONSIDER THE SIGNS

Alternatively, look at the signs of the product and the sum.

- The product is negative, so p and q have different signs. Their sum is the *difference* in their absolute values.

- The sum is positive, so the one with the greater absolute value is positive.

Now look for the absolute values of p and q: two positive numbers whose product is 30 and whose difference is 7. That's 3 and 10. The one with the greater absolute value is positive, the other one is negative, so the factorization is $(x + 10)(x - 3)$.

KEY POINTS

Consider the Signs When factoring $x^2 + bx + c$, the signs of b (the sum of p and q) and c (the product of p and q) can tell you a lot about p and q.

- If c is positive, then p and q have the same sign—both have the same sign as b.

- If c is negative, then p and q have opposite signs. The one whose absolute value is greater has the same sign as b. The other one has the opposite sign.

EXAMPLE: Signs of b and c

Replace each □ with a plus sign or a minus sign.

a. $x^2 - 9x + 14 = (x \;\square\; 2)(x \;\square\; 7)$

b. $x^2 - 5x - 6 = (x \;\square\; 6)(x \;\square\; 1)$

c. $x^2 + 5x + 6 = (x \;\square\; 3)(x \;\square\; 2)$

d. $x^2 + 10x - 39 = (x \;\square\; 3)(x \;\square\; 13)$

SOLUTION

a. The constant term (14) is positive, so both 2 and 7 get the sign of the cross-term ($-9x$): $x^2 - 9x + 14 = (x - 2)(x - 7)$.

b. The constant term (-6) is negative, so the larger of 6 and 1 gets the sign of the cross-term ($-5x$): $x^2 - 5x - 6 = (x - 6)(x + 1)$.

c. The constant term (6) is positive, so both 3 and 2 get the sign of the cross-term ($5x$): $x^2 + 5x + 6 = (x + 2)(x + 3)$.

d. The constant term (-39) is negative, so the larger of 3 and 13 gets the sign of the cross-term ($10x$):

$$x^2 + 10x - 39 = (x - 3)(x + 13).$$

FACTORING $ax^2 + bx + c$

Factoring a trinomial whose lead coefficient is not 1 is a little more involved than the examples with $x^2 + bx + c$ that we've been looking at so far. The methods are similar, though: you still use trial and error to find two signed numbers with a specific sum and a specific product.

Before you plunge into this section, check that factoring a general quadratic like $ax^2 + bx + c$ is something that you need to know how to do. The *quadratic formula*, which we'll look at in Chapter 6, essentially allows you to bypass this trial-and-error method, so some teachers don't bother with factoring $ax^2 + bx + c$ if a is not 1. Many others do, however, so check with yours.

The method presented here is sometimes known as the *key number method*. As usual, the numbers will be smaller and the computations easier if you pull out any common factors first.

Here are the steps for factoring $ax^2 + bx + c$. We'll show how you can use them to factor $3x^2 + 5x - 2$ as we go along.

1. Multiply a by c to get the signed number ac. Some textbooks call ac the *key number*.

   EXAMPLE:
   Factor $3x^2 + 5x - 2$.

   SOLUTION
   Here a is 3 and c is –2, so ac is –6.

2. Find two signed numbers p and q whose product is ac and whose (signed) sum is b.

   We need to find p and q whose product is –6 and whose sum is 5. With trial and error, we find that 6 and –1 work.

3. Express $ax^2 + bx + c$ as $ax^2 + px + qx + c$. It makes sense to do this since b is the sum of p and q.

$$3x^2 + 5x - 2 = 3x^2 + 6x - x - 2$$

4. Group $(ax^2 + px) + (qx + c)$ and factor by pulling out common factors.

$$
\begin{aligned}
3x^2 + 6x - 2 &= (3x^2 + 6x) + (-x - 2) \\
&= 3x(x + 2) + (-1)(x + 2) \\
&= (3x - 1)(x + 2)
\end{aligned}
$$

Check by multiplying that $(3x - 1)(x + 2) = 3x^2 + 5x - 2$.

Try these examples on your own before reading through the solutions.

EXAMPLE: Factor $ax^2 - bx - c$
Factor $8x^2 - 2x - 3$.

SOLUTION
The product ac is $8 \cdot (-3) = -24$. Now we need two numbers that multiply to -24 and add to -2. Since -24 is negative, one of the numbers must be positive and the other negative. Since -2 is negative, the number with the greater absolute value must be negative.

The only pairs that fit these conditions are 1 and -24, 2 and -12, 3 and -8, and 4 and -6. Only 4 and -6 add to -2.

Re-express, group, and factor:

$$
\begin{aligned}
8x^2 - 2x - 3 &= 8x^2 + 4x - 6x - 3 \\
&= (8x^2 + 4x) + (-6x - 3) \\
&= 4x(2x + 1) + (-3)(2x + 1) \\
&= (4x - 3)(2x + 1)
\end{aligned}
$$

Check: $(4x - 3)(2x + 1) = 8x^2 + 4x - 6x - 3 = 8x^2 - 2x - 3$.

Factoring Polynomials • 221

EXAMPLE: Factor $ax^2 - bx + c$

Factor $15x^2 - 34x + 15$.

SOLUTION

The product of the first and last coefficient is 225. Now we need two numbers that multiply to 225 and add to –34. Since 225 is positive, both numbers have the same sign. Since –34 is negative, both numbers must be negative.

The pairs of numbers that fit these conditions are –1 and –225, –3 and –75, –5 and –45, –9 and –25, and –15 and –15. Of these, only –9 and –25 add up to –34.

Re-express, group, and factor:

$$
\begin{aligned}
15x^2 - 34x + 15 &= 15x^2 - 9x - 25x + 15 \\
&= (15x^2 - 9x) + (-25x + 15) \\
&= 3x(5x - 3) + (-5)(5x - 3) \\
&= (3x - 5)(5x - 3)
\end{aligned}
$$

Check: $(3x - 5)(5x - 3) = 15x^2 - 9x - 25x + 15 = 15x^2 - 34x + 15$.

Summary

Polynomials

- A *monomial* is a product of numbers and variables.
- A *polynomial* is a sum of monomials.

Degrees

- The *degree* of a monomial is the sum of the exponents on each variable. The degree of a polynomial is the greatest degree of any of its monomial terms.

Like terms

- *Like terms* are monomials with the same variables to the same degrees. Like terms may be added or subtracted into one term.

Multiplying polynomials

- To multiply two polynomials, multiply each term of the first by each term of the second, using the distributive property twice. In particular,

$$(a + b)(c + d) = ac + ad + bc + bd$$

First, Outer, Inner, Last.

Long division of polynomials

- Polynomials in one variable can be long-divided, much like integers.

Factoring polynomials

1. Pull out common factors from every term. Always do this first.
2. Look for special products: $a^2 \pm 2ab + b^2$ or $a^2 - b^2$.

3. Try factoring by grouping. The polynomial must have an even number of terms and an even number of negative coefficients.

4. If the polynomial is in $x^2 + bx + c$ form, look for two signed numbers p and q whose sum is b and whose product is c. Then $x^2 + bx + c$ factors as $(x + p)(x + q)$.

5. If the polynomial is in $ax^2 + bx + c$ form, look for two signed numbers p and q whose sum is b and whose product is ac. Then factor $ax^2 + px + qx + c$ by grouping.

   Not every polynomial can be factored.

Sample Test Questions

Answers to these questions begin on page 396.

1. Say whether each of these expressions is a polynomial. If it is, give its degree.

   A. $4x^2 - 7x + 12$

   B. $\dfrac{z^2 + 1}{z - 2}$

   C. $-11 + \sqrt{3}$

   D. $1 - y + y^2 - y^3 + y^4 - y^5$

   E. $2x - \sqrt{x}$

   F. $x^2 y - xy^2 + 3xy$

   G. $x + y + z$

2. Classify each polynomial according to the number of terms (monomial, binomial, or trinomial) and the degree (constant, linear, quadratic, cubic, quartic, or quintic). For example, $x^2 + 2x + 4$ is a quadratic trinomial.

   A. $x + 4$

   B. $-6y^5$

   C. $x^3 + 2x + 1$

   D. $z^4 - z^2$

3. Add or subtract, as indicated.

   A. $(x^2 - 3x + 1) + (4x + 6)$

   B. $(3x^2 + 2x - 4) - (x^2 - 4x + 5)$

   C. $2(x^2 + 5) + (x - 3)$

   D. $(-x^3 + x^2 - 3) - 2(x^3 - 2x + 4)$

4. Multiply, as indicated.

   A. $2x(x^2 - 4x + 7)$

   B. $-xy(x^2 + y^2 - x - y)$

   C. $(x + 3)(x + 5)$

D. $(x - 2)(x + 6)$

E. $(3x - 2)(2x - 5)$

F. $(-x - 2y)(3x - 7y)$

G. $(x - 3)(x^3 + 5x^2 - 2x - 4)$

5. Use long division to find the quotient and remainder when $2x^3 + 7x^2 - 9x + 14$ is divided by $x + 5$.

6. What is $x^3 - 7x - 16$ divided by $x - 4$? If you know it, use synthetic division.

7. Factor each polynomial by pulling out a common monomial factor.

A. $12x - 8$

B. $z^7 - z^5 + z^2$

C. $-2x^3y + 10x^2y^2 - 6xy^3$

8. Factor each of these polynomials by grouping.

A. $y^3 - 4y^2 + 2y - 8$

B. $2xy + 6x - y - 3$

C. $4x^2 + 9x + 28x + 63$

D. $3x^2y - 3xy^2 - 6x^2 + 6xy$

E. $ab - 2b + 3ac - 6c + 7ad - 14d$

9. Factor each polynomial. *Hint:* Look for special products.

A. $x^2 - 36$

B. $100 - y^2$

C. $x^2 - 16x + 64$

D. $9x^2 + 6x + 1$

E. $3 - 27x^2$

F. $-2z^2 + 44z - 242$

G. $25x^2 + 30x + 9$

H. $x^4 - 14x^2 + 49$

10. Factor each polynomial.

    A. $x^2 + 8x + 7$

    B. $x^2 - 7x + 10$

    C. $x^2 - 5x - 6$

    D. $x^2 + x - 6$

    E. $3x^2 + 12x - 36$

11. Factor each polynomial.

    A. $2x^2 + x - 1$

    B. $3x^2 + 13x + 4$

    C. $2x^2 + x - 21$

    D. $3x^2 - 17x + 10$

    E. $8x^2 - 2x - 3$

12. Factor each polynomial or say that the polynomial does not factor.

    A. $x^4 - 81$

    B. $(x - 1)^2 - 4$

    C. $y^2 - 6y + 9 - 25z^2$

    D. $x^2 + 9$

    E. $x^3 + x^2 - x - 1$

    F. $ab + b - 2a + 2$

    G. $x^4 + 4$ *Hint:* Re-express as $(x^4 + 4 + 4x^2) - 4x^2$.

13. The sum of a and another number is 21. The product of the two numbers is 80. Translate these statements to an equation in terms of a.

14. Emma and Ruby, two Labrador retrievers, are fetching sticks thrown by their owners in the park. Emma fetches x sticks per hour for y hours. Ruby fetches 6 more sticks per hour than Emma, but because her owner has to go home early, Ruby spends 1 hour less than Emma fetching sticks. During their time in the park, Ruby fetches a total of 3 more sticks than Emma.

A. How many sticks does Emma fetch that day? Express your answer as a polynomial in x and y.

B. How many sticks does Ruby fetch that day? Express your answer as a polynomial in x and y.

C. Write down an equation that relates x and y. Your equation should be linear.

D. If Emma spends 3 hours fetching sticks, how many sticks does Ruby fetch per hour? Use your equation from part c.

Quadratic Equations

6

Overview

A linear equation in one variable is a sum or difference of xs and numbers, so it can always be made to look like $ax + b = 0$. Linear equations have straight-line graphs—nice and simple, but a little boring. In this chapter we turn it up a notch and look at *quadratic equations*, which are sums and differences of x^2s, xs, and numbers. Quadratic equations can always be made to look like $ax^2 + bx + c = 0$. Their graphs are curvy shapes called *parabolas*.

Quadratic equations are both versatile and approachable, tricky but not too tricky. Once you know the keys to unlock their secrets, they're completely manageable. The *quadratic formula* is the most important of these keys.

Polynomial Equations and Roots

POLYNOMIAL EQUATIONS

A *polynomial equation* is simply an equation that involves only polynomial terms—that is, real numbers, variables, and their sums and products. The following are all examples of polynomial equations:

$$3x^2 - 4x = 5 \qquad z^5 = 1 \qquad a^2b - ab^2 = 3ab$$

One of the big pursuits of algebra is looking for solutions to polynomial equations in one variable; in other words, finding all the values of the variable that make the equation into a true statement. We already discussed how to solve equations such as $2x + 7 = 6$: these are *linear* equations in one variable (Chapter 1). The bulk of this chapter is dedicated to solving equations like $3x^2 - 4x = 5$: *quadratic* equations in one variable.

ROOTS OF POLYNOMIALS

In order to simplify the problem of finding solutions to polynomial equations in one variable, we often rearrange the equation to get a polynomial in one variable on one side and zero on the other side. So instead of looking for solutions to $3x^2 - 4x = 5$, we can look for solutions to $3x^2 - 4x - 5 = 0$, and instead of solving $z^5 = 1$, we try to solve $z^5 - 1 = 0$. A solution to an equation in which a polynomial equals zero is called a **root** of the polynomial and sometimes also a **zero** of the polynomial.

EXAMPLE: Checking roots

a. Is -4 a root of $x^3 + 64$?

b. Is 4 a root of $x^3 + 64$?

c. Is -3 a root of $x^2 - 2x - 15$?

d. Is -5 a root of $x^2 - 2x - 15$?

e. Is $\dfrac{2}{3}$ a root of $3x^2 + x - 2$?

SOLUTION
To check if a number is a root of a polynomial, plug the number into the polynomial and see if the expression evaluates to 0.

a. Yes: $(-4)^3 + 64 = -64 + 64 = 0$, so -4 is a solution to $x^3 + 64 = 0$—that is, a root of $x^3 + 64$.

b. No: $4^3 + 64 = 64 + 64 \neq 0$, so 4 *is not* a root of $x^3 + 64$.

c. Yes: $(-3)^2 - 2(-3) - 15 = 9 + 6 - 15 = 0$, so -3 *is* a root of $x^2 - 2x - 15$.

d. No: $(-5)^2 - 2(-5) - 15 = 25 + 10 - 15 \neq 0$, so -5 *is not* a root of $x^2 - 2x - 15$.

e. Yes: $3\left(\dfrac{2}{3}\right)^2 + \dfrac{2}{3} - 2 = 3\left(\dfrac{4}{9}\right) + \dfrac{2}{3} - 2 = 0$, so $\dfrac{2}{3}$ *is* a root of $3x^2 + x - 2$.

QUADRATIC EQUATIONS

A **quadratic equation** is an equation in one variable that involves polynomial terms of degree 2 and less. (The *degree* of a polynomial term is the exponent on the variable; the degree of $3x^2$ is 2 and the degree of $-4z$ is 1.)

For example, $3x^2 - 4x + 7 = 0$ is a quadratic equation; so is $z^2 + 3 = -4z$. But $4x^3 + x^2 = 1$ is not a quadratic because it involves a term of degree 3. Neither is $\frac{3}{x} + x^2 = 8$: it has a variable in the denominator and so is not even a polynomial.

By moving terms to one side of the equation, a quadratic equation in x can always be written in the form

$$ax^2 + bx + c = 0$$

where a, b, and c are real numbers and a is not 0. (If $a = 0$, then there is no x^2 term, so the equation is *linear* rather than quadratic.)

EXAMPLE: Is equation quadratic?

For each equation, determine whether it is a quadratic equation in x. If it is, express it in $ax^2 + bx + c = 0$ form.

a. $x^7 + x^2 = 1$

b. $4x = x^2 + 3$

c. $-\frac{7}{2}x^2 = 3x$

d. $x^2 = \sqrt{x} + \frac{2}{x+1}$

e. $3x - x^2 = 6 - x^2$

SOLUTION

a. Not a quadratic equation: the term x^7 has degree 7, so this polynomial does not have degree 2.

b. This is a quadratic equation. Moving the $4x$ on the left side of the equation to the right side (by adding $-4x$ to both sides of the equation) gives $0 = x^2 - 4x + 3$, or, equivalently, $x^2 - 4x + 3 = 0$. Note that more than one answer is possible here: if you move the nonzero terms to the left side, you'll get $-x^2 + 4x - 3 = 0$, which is an equivalent quadratic equation.

c. This is a quadratic equation. Subtracting $3x$ from both sides (which is the same thing as adding $-3x$ to both sides) gives $-\frac{7}{2}x^2 - 3x = 0$, which is in $ax^2 + bx + c = 0$ form with $c = 0$.

d. Not a quadratic equation. There is a term with the variable under a radical sign and a term with the variable in the denominator. Either is enough to disqualify the equation from being polynomial.

e. This equation looks quadratic, but in fact it is not. Moving all the terms to, say, the left side of the equation gives $3x - x^2 - 6 + x^2 = 0$, which simplifies to $3x - 6 = 0$. The quadratic terms cancel, leaving a linear polynomial equation.

Solving Quadratic Equations

There are several common algebraic ways of finding solutions to quadratic equations. (Keep in mind that the solutions to $ax^2 + bx + c = 0$ are, by definition, the roots of $ax^2 + bx + c$, so we'll be using these phrases interchangeably.)

- **Extracting roots:** Simple quadratics that have no linear term (that is, quadratics that look like $x^2 = d$ or like $ax^2 + c = 0$) can be solved directly by taking, or *extracting*, square roots.

- **Factoring:** Some quadratic equations can be solved by *factoring*. This is a powerful method—but it only works for quadratics that can be factored. In practice, it is only used to solve quadratics that are easy to factor.

- **Completing the square:** Any quadratic can be solved by *completing the square*. Completing the square is a useful technique that comes up again and again when working with polynomials. However, many people prefer solving quadratic equations with the quadratic formula.

- **Quadratic formula:** The *quadratic formula*, derived by completing the square on the general equation $ax^2 + bx + c = 0$, gives an exact formula for the solutions in terms of the coefficients a, b, and c. This popular formula is widely used on quadratic equations arising from real-life problems, which are, in general, difficult to factor.

SOLVING BY EXTRACTING SQUARE ROOTS

Solving $x^2 = d$ for positive d

What are all the solutions to $x^2 = 4$?

Each positive number d has exactly two square roots, $\sqrt{d}$ and $-\sqrt{d}$. Since $\sqrt{4} = 2$, there are exactly two solutions to $x^2 = 4$: $x = 2$ and $x = -2$. You can check that both work: $2^2 = 4$ and $(-2)^2 = 4$.

We can use the same reasoning on any quadratic equation of the form $x^2 = d$ where d is a positive number:

> **Two Solutions**
>
> If $d > 0$, then the equation
> $$x^2 = d$$
> has exactly two solutions:
> $$x = \sqrt{d} \text{ and } x = -\sqrt{d}$$

EXAMPLE: Solve $x^2 = d$

Find all solutions to $x^2 = 12$.

SOLUTION

Since 12 is positive, the equation has exactly two solutions: $x = \sqrt{12}$ and $x = -\sqrt{12}$. For completeness, we can simplify $\sqrt{12}$ a little bit:

$$\sqrt{12} = \sqrt{4 \cdot 3} = 2\sqrt{3}$$

So $x = 2\sqrt{3}$ and $x = -2\sqrt{3}$ are the only solutions. This is sometimes written as $x = \pm 2\sqrt{3}$.

You can think of the solution as taking the square root of both sides of $x^2 = 12$. Since squaring erases the distinction between positive and negative numbers, you must introduce that distinction back when you take square roots: $x^2 = 12$ is equivalent $x = \pm\sqrt{12}$.

Now try two more complicated examples. See if you can work them out before looking at the solutions.

EXAMPLE: Solve $ax^2 = d$

Find all solutions to $25x^2 = 9$.

SOLUTION

To solve this equation, we isolate the x^2 term on the left side by dividing both sides by 25:

$$\frac{25}{25}x^2 = \frac{9}{25}$$

$$x^2 = \frac{9}{25}$$

Since $\frac{9}{25}$ is positive, we can take square roots to get

$$x = \pm\sqrt{\frac{9}{25}}$$

which simplifies to

$$x = \pm\frac{\sqrt{9}}{\sqrt{25}} = \pm\frac{3}{5}$$

So, the solutions are $\frac{3}{5}$ and $-\frac{3}{5}$.

EXAMPLE: Solve $ax^2 - d = 0$

Find all solutions to $12x^2 - 225 = 0$.

SOLUTION

To solve this equation, we first isolate the x^2 term on one side and the number term on the other side:

$$12x^2 - 225 = 0$$
$$12x^2 - 225 + 225 = 0 + 225$$
$$\frac{12}{12}x^2 = \frac{225}{12} = \frac{\overset{75}{\cancel{225}}}{\underset{4}{\cancel{12}}}$$
$$x^2 = \frac{75}{4}$$

Now, extract square roots and simplify:

$$x = \pm\sqrt{\frac{75}{4}}$$
$$= \pm\frac{\sqrt{25 \cdot 3}}{\sqrt{4}}$$
$$= \pm\frac{\sqrt{25}\sqrt{3}}{\sqrt{4}}$$
$$= \pm\frac{5\sqrt{3}}{2}$$

Solving $x^2 = d$ for negative d or d = 0

So far, we've looked at quadratic equations that can be reduced to $x^2 = d$ with d positive. What happens if d is negative?

EXAMPLE: Solve $x^2 = -d$

What are all solutions to $x^2 = -4$?

SOLUTION

Since the square of any real number is either positive or zero, this equation has no real-number solutions. There is no real value of x that makes this equation true.

The equation $x^2 + 4 = 0$ also has no solutions. That's because $x^2 + 4 = 0$ and $x^2 = -4$ are equivalent equations, so they have the same solutions.

The same thing will happen whenever you try to solve the equation $x^2 = d$ for any negative d. No real number has a negative square, so $x^2 = d$ has no solutions if d is negative.

To complete the story, let's look at $x^2 = d$ for $d = 0$.

EXAMPLE: Solve $x^2 = 0$
Find all solutions to $x^2 = 0$.

SOLUTION
Zero has only one square root—itself. So $x = 0$ is the only solution.

KEY POINTS

$x^2 = d$ Quadratic equations of the form $x^2 = d$ come in three different types.

1. If d is positive, $x^2 = d$ has exactly two solutions, $\pm\sqrt{d}$.

2. If d is negative, $x^2 = d$ has no solutions.

3. If d is zero, there is exactly one solution: $x = 0$.

We can now solve any equation that looks like $x^2 = d$.
   Additionally, we can solve any equation that looks like:

- $x^2 + c = 0$ (just move the c over to the other side)

- $ax^2 = d$ (divide both sides by a to get rid of it)

- $ax^2 + c = 0$ (do both)

Like $x^2 = d$ equations, Equations in the form $ax^2 + c = 0$ come in three types: those with two real solutions, those with no real solutions and those with one real solution. It turns out that *all* quadratic equations can be classified in the same way: about half have two real roots, about half have no real roots, and some have exactly one real root.

SOLVING BY FACTORING

Zero Product Property

The factoring method of solving quadratic equations depends on the zero product property:

> **Zero Product Property**
>
> If a and b are real numbers and $ab = 0$,
> then $a = 0$ and/or $b = 0$ or both.

The zero product property is another way of saying that the product of two nonzero numbers is never zero.

The Zero Product Property in Action

Take a look at how the zero product property can be applied to solve equations.

EXAMPLE: Zero product property I
Find all the solutions to $(x - 1)(x - 2) = 0$.

SOLUTION
By the zero product property, if $(x - 1)(x - 2) = 0$, then either $x - 1 = 0$ or $x - 2 = 0$:

$$
\begin{array}{ccc}
x - 1 = 0 & \text{or} & x - 2 = 0 \\
\downarrow & & \downarrow \\
x = 1 & \text{or} & x = 2
\end{array}
$$

So if $(x - 1)(x - 2) = 0$, then either $x = 1$ or $x = 2$. In other words, $(x - 1)(x - 2) = 0$ has exactly two solutions: $x = 1$ and $x = 2$.

Try the next two examples on your own before looking at the solutions.

EXAMPLE: Zero product property II

Find all solutions to $x(x + 3) = 0$.

SOLUTION

By the zero product property, if $x(x + 3) = 0$, then either $x = 0$ or $x + 3 = 0$:

$$x = 0 \quad \text{or} \quad x + 3 = 0$$
$$\downarrow$$
$$x = -3$$

So $x(x + 3) = 0$ has two solutions: $x = 0$ and $x = -3$.

It might be tempting to attack this problem by multiplying out first to get $x^2 + 3x = 0$. This isn't an unreasonable first step, but this section is all about solving by factoring: first factor, then use the zero product property on the factored form. The equation $x(x + 3) = 0$ is already factored, so we can start with the zero product property right off.

EXAMPLE: Zero product property III

Find all solutions to $(2x - 1)(3x + 4) = 0$.

SOLUTION

By the zero product property, if $(2x - 1)(3x + 4) = 0$, then either $2x - 1 = 0$ or $3x + 4 = 0$:

$$2x - 1 = 0 \quad \text{or} \quad 3x + 4 = 0$$
$$\downarrow \qquad\qquad\qquad \downarrow$$
$$2x = 1 \quad \text{or} \quad 3x = -4$$
$$\downarrow \qquad\qquad\qquad \downarrow$$
$$x = \frac{1}{2} \quad \text{or} \quad x = -\frac{4}{3}$$

So $(2x - 1)(3x + 4) = 0$ has two solutions: $x = \frac{1}{2}$ and $x = -\frac{4}{3}$.

The Zero Product Property and Factoring

What if the equation you're trying to solve isn't presented as a product of two linear terms?

That's where factoring comes in. If you can factor $ax^2 + bx + c$ into a product of two linear terms, then you can solve the equation $ax^2 + bx + c = 0$ using the zero product property, just as we've been doing. Refer to Chapter 5 to review how to factor quadratic polynomials.

EXAMPLE: Solve $x^2 + bx + c = 0$ by factoring

Solve the equation $x^2 + 2x - 8 = 0$.

SOLUTION

First factor $x^2 + 2x - 8$. This means that we need to find two signed numbers whose sum is 2 and whose product is –8. Since the product is negative, one of the numbers is positive and the other is negative. The pairs of numbers whose product is –8 are:

$$-1 \text{ and } 8, \qquad \text{and} \qquad -2 \text{ and } 4,$$
$$-4 \text{ and } 2, \qquad\qquad -8 \text{ and } 1$$

The only pair whose sum is 2 is –2 and 4.

Check that $x^2 + 2x - 8 = (x + 4)(x - 2)$ by multiplying:

$$(x + 4)(x - 2) = x^2 + 4x - 2x - 8 = x^2 + 2x - 8$$

so the factorization works.

Now use the zero product property to solve $(x + 4)(x - 2) = 0$. If $(x + 4)(x - 2) = 0$, then either $x + 4 = 0$ or $x - 2 = 0$:

$$x + 4 = 0 \qquad \text{or} \qquad x - 2 = 0$$
$$\downarrow \qquad\qquad\qquad \downarrow$$
$$x = -4 \qquad \text{or} \qquad x = 2$$

So $x^2 + 2x - 8 = 0$ has two solutions, $x = -4$ and $x = 2$.

Try the next examples on your own before reading through the solutions.

EXAMPLE: Solve $ax^2 - bx - c = 0$ by factoring
Find all solutions to $3x^2 - 20x - 7 = 0$.

SOLUTION
To solve $3x^2 - 20x - 7 = 0$, we first factor $3x^2 - 20x - 7$.

To factor the polynomial, we find the product of the first and last coefficients: $3 \cdot (-7) = -21$. Next we look for two signed numbers whose product is -21 and whose sum is -20, which is the middle coefficient. The pairs of numbers whose product is -21 are -1 and 21, -3 and 7, -7 and 3, and -21 and 1. Only -21 and 1 add up to -20.

Rewrite $3x^2 - 20x - 7$ as $3x^2 - 21x + x - 7$ and factor by grouping:

$$3x^2 - 21x + x - 7 = 3x(x - 7) + 1(x - 7)$$
$$= (3x + 1)(x - 7)$$

Now we can use the zero product property. If $(3x + 1)(x - 7) = 0$, then $3x + 1 = 0$ or $x - 7 = 0$:

$$3x + 1 = 0 \qquad \text{or} \qquad x - 7 = 0$$
$$\downarrow \qquad\qquad\qquad \downarrow$$
$$x = -\frac{1}{3} \qquad \text{or} \qquad x = 7$$

So the two solutions to $3x^2 - 20x - 7 = 0$ are $x = -\frac{1}{3}$ and $x = 7$.

EXAMPLE: Find roots of $ax^2 + bx + c$

Find all roots of the polynomial $18x^2 + 24x + 8$.

SOLUTION

The roots of $18x^2 + 24x + 8$ are the solutions to $18x^2 + 24x + 8 = 0$. To find them, we need to factor $18x^2 + 24x + 8$.

To factor $18x^2 + 24x + 8$, first note that we can pull out a factor common to all the coefficients:

$$18x^2 + 24x + 8 = 2(9x^2 + 12x + 4)$$

Next, we must factor $9x^2 + 12x + 4$. The product of the first and last coefficients is $9 \cdot 4 = 36$. We look for two numbers whose product is 36 and whose sum is 12. Trial and error shows that 6 and 6 work. Factor by grouping:

$$\begin{aligned} 9x^2 + 12x + 4 &= 9x^2 + 6x + 6x + 4 \\ &= 3x(3x + 2) + 2(3x + 2) \\ &= (3x + 2)(3x + 2) \end{aligned}$$

So $18x^2 + 24x + 8$ factors as $2(3x + 2)(3x + 2) = 2(3x + 2)^2$.

By the zero product property, if $2(3x + 2)(3x + 2) = 0$, then $3x + 2 = 0$ or $3x + 2 = 0$ (or both—the two cases are redundant in this situation). So $3x + 2$ must equal 0, which means that $x = -\dfrac{2}{3}$.

Since $x = -\dfrac{2}{3}$ is the only solution to the equation $18x^2 + 24x + 8 = 0$, it's also the only root of the polynomial $18x^2 + 24x + 8$.

OPTIONS

Extracting Roots or Factoring? Take a look at the following example:

EXAMPLE:
Find all solutions to the equation $x^2 - 100 = 0$.

We can solve $x^2 - 100 = 0$ by extracting square roots. But since $x^2 - 100$ is a difference of squares, which is easy to factor, we can find solutions that way. Let's try both ways.

SOLUTION: Option one: Extracting roots
We can rewrite $x^2 - 100 = 0$ by moving the -100 to the right side:

$$x^2 = 100$$

Taking the square root of both sides gives

$$x = \pm\sqrt{100} = \pm 10$$

So the solutions are $x = 10$ and $x = -10$.

SOLUTION: Option two: Factoring
$x^2 - 100$ is a difference of squares, so it isn't difficult to factor:

$$x^2 - 100 = x^2 - 10^2$$
$$= (x + 10)(x - 10)$$

By the zero product property, if $x^2 - 10 = (x + 10)(x - 10) = 0$, then $x + 10 = 0$ or $x - 10 = 0$, which is equivalent to $x = -10$ or $x = 10$.

The two methods give us the same solutions, as expected. The first method is simpler, but the second is somewhat general. If you happen not to notice that the first method will work, the second method is still a great option.

SOLVING BY COMPLETING THE SQUARE

The drawback of the factoring method of solution is that it only works when the polynomial in question is easy to factor.

Completing the square, a generalization of the direct-root-extraction method (see page 234), along with its cousin the *quadratic formula* (next section), is a much more versatile method than factoring.

Warm-up

Earlier in this chapter, we looked at how to solve equations such as $x^2 = d$ and $ax^2 = d$ by taking square roots of both sides. You can also use the same method to solve any equation in the form $a(x + p)^2 = d$. Take a look at an example to see how this works.

EXAMPLE: Solve $(x + p)^2 = d$
Solve $(x - 3)^2 = 49$.

SOLUTION
Just as if we were dealing with $x^2 = 49$, we can take the square root of both sides. Pay attention to how the ± sign is treated. Since

$$(x - 3)^2 = 49$$

we know that

$$x - 3 = \pm\sqrt{49}$$
$$= \pm 7$$

And now that we know that $x - 3 = 7$ or $x - 3 = -7$, we can examine each case individually:

$$
\begin{array}{ccc}
x - 3 = 7 & \text{or} & x - 3 = -7 \\
\downarrow & & \downarrow \\
x = 10 & \text{or} & x = -4
\end{array}
$$

So the two solutions to $(x - 3)^2 = 49$ are $x = 10$ and $x = -4$.
  You can plug each in to check that it works: $(10 - 3)^2 = 7^2 = 49$ and $(-4 - 3)^2 = (-7)^2 = 49$.

CHAPTER 6
QUADRATIC EQUATIONS

The next example is only a little more involved. Try to work it out on your own before reading the solution.

EXAMPLE: Solve $a(x + p)^2 = d$

Find the solutions to $4(x + 1)^2 = 25$.

SOLUTION

We can divide the equation through by 4 to eliminate the coefficient on the left and then take square roots of both sides, just as if we were dealing with $4x^2 = 25$.

Divide by 4:

$$4(x + 1)^2 = 25$$

$$(x + 1)^2 = \frac{25}{4}$$

Take the square root of both sides:

$$x + 1 = \pm\sqrt{\frac{25}{4}}$$

$$= \pm\frac{5}{2}$$

And now treat the two cases individually:

$$x + 1 = \frac{5}{2} \qquad \text{or} \qquad x + 1 = -\frac{5}{2}$$

$$\downarrow \qquad\qquad\qquad\qquad \downarrow$$

$$x = \frac{5}{2} - 1 \qquad \text{or} \qquad x = -\frac{5}{2} - 1$$

$$\downarrow \qquad\qquad\qquad\qquad \downarrow$$

$$x = \frac{3}{2} \qquad \text{or} \qquad x = -\frac{7}{2}$$

So the two solutions to $4(x + 1)^2 = 25$ are $x = \frac{3}{2}$ and $-\frac{7}{2}$. Plug each value into the original equation to check that it works.

Both of these examples had two solutions—but we already know that this is not always the case. Try your hand at this last example.

EXAMPLE: Solve $(ax + p)^2 = -d$
Solve $(5x - 2)^2 = -9$.

SOLUTION
If the equation $(5x - 2)^2 = -9$ is to be a true statement, then the square of some number (namely, of $5x - 2$) has to be a negative number (namely, -9). But the square of any number is nonnegative; there can be no values of $5x - 2$ with square -9. Therefore no real values of x that make the equation true.
    So $(5x - 2)^2 = -9$ has no solutions.

In short, we can solve any quadratic equation in the form $a(x + p)^2 = d$, even if it has no solutions.

Unfortunately, many quadratic equations come in the form $ax^2 + bx + c = 0$, not in the form $a(x + p)^2 = d$. Fortunately, *completing the square* is a procedure that helps you re-express the equation $ax^2 + bx + c = 0$ as $a(x + p)^2 = d$.

Perfect Squares

From Chapter 5, you know that $(x + p)^2$ is

$$x^2 + 2px + p^2$$

Look at this expression closely. If you know $2p$, the x coefficient, you can find p by halving it. And then you can find p^2 by squaring p.

Practice with this example.

EXAMPLE: Complete perfect squares

Fill in the missing terms for each of these perfect squares:

a. $(x + 3)^2 = x^2 + 6x + \underline{}$

b. $(x - 2)^2 = x^2 - 4x + \underline{}$

c. $(x \underline{})^2 = x^2 + 14x + \underline{}$

d. $(x \underline{})^2 = x^2 - 20x + \underline{}$

e. $(x \underline{})^2 = x^2 - 9x + \underline{}$

SOLUTION

a. The middle term is $6x$; half its coefficient is 3, which is con-sistent with the $(x + 3)^2$ on the left side. Squaring 3 gives 9, so the answer is

$$(x + 3)^2 = x^2 + 6x + \underline{9}$$

b. The middle term is $-4x$; half its coefficient is -2, which is consistent with the $(x - 2)^2$ on the left. Squaring -2 gives 4:

$$(x - 2)^2 = x^2 - 4x + \underline{4}$$

c. The middle coefficient is 14, so the left side must be $(x \underline{+ 7})^2$. Since 7^2 is 49, that's the constant term on the right-hand side:

$$(x \underline{+ 7})^2 = x^2 + 14x + \underline{49}$$

d. The middle coefficient is -20, so the left side is $(x \underline{- 10})^2$. Square -10 to complete the right-hand side:

$$(x \underline{- 10})^2 = x^2 - 20x + \underline{100}$$

e. The middle coefficient is -9; half of that is $-\dfrac{9}{2}$, so the left-hand side is $\left(x - \dfrac{9}{2}\right)^2$. The square of $-\dfrac{9}{2}$ is $\dfrac{81}{4}$, so that completes the right side:

$$\left(x - \frac{9}{2}\right)^2 = x^2 - 9x + \frac{81}{4}$$

Polynomials are expressions obtained by adding, subtracting, and multiplying real numbers and one or several variables. Usually the variables are arranged alphabetically.

—Expressions connected by + or − signs are called **terms**. **Ex:** The polynomial $2x^3y - 7x$ has two **terms.**

—The **coefficient** of a term is the real number (non-variable) part.

—Two terms are sometimes called **like terms** if the power of each variable in the terms is the same. **Ex:** $7y^6x$ and $yxy^5 = xy^6$ are like terms. $2x^8$ and $16xy^7$ are not. Like terms can be added or subtracted into a single term.

—The **degree** of a term is the sum of the powers of each variable in the term. **Ex:** $2x^8$ and $16xy^6z$ both have degree eight.

—The degree of a polynomial is the highest degree of any of its terms.

—In a polynomial in one variable, the term with the highest degree is called the **leading term,** and its coefficient is the **leading coefficient.**

CLASSIFICATION OF POLYNOMIALS

By degree: Ex: $2x^5y - 4x^3y^3 +$ both sixth-degree polynomials
Special names for polynomials

By degree:	By number
degree 1: **linear**	1 term
degree 2: **quadratic**	2 terms
degree 3: **cubic**	3 terms
degree 4: **quartic**	
degree 5: **quintic**	

A linear equation in two variables (say $ax + by = c$ with a and c not both zero) has infinitely many ordered pair (x, y) **solutions**—real values of x and y that make the equation true. Two simultaneous linear equations in two variables will have:

—Exactly **one solution** if their graphs intersect—the most common scenario.

—**No solutions** if the graphs of the two equations are parallel.

—**Infinitely many solutions** if their graphs coincide.

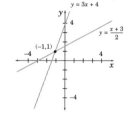

SOLVING BY GRAPHING: TWO VARIABLES

Graph both equations on the same Cartesian plane. The intersection of the graph gives the simultaneous solutions. (Since points on each graph correspond to solutions to the appropriate equation, points on *both* graphs are solutions to both equations.)

—Sometimes, the exact solution can be determined from the graph; other times the graph gives an estimate only. Plug in and check.

—If the lines intersect in exactly one point (most cases), the intersection is the unique solution to the system (*See graph at top of next column*).

—If the lines are parallel, they do not intersect; the system has no solutions. *Parallel lines have the same slope;* if the slope is not the same, the lines will intersect.

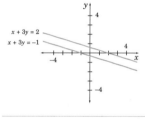

—If the lines coincide, there are infinitely many solutions. Effectively, the two equations convey the same information (*See graph at top of next column*).

SOLVING BY SUBSTITUTION: TWO VARIABLES

—Use one equation to solve for one variable (say, y) in terms of the other (x): isolate y on one side of the equation.

—Plug the expression for y into the other equation.

—Solve the resulting one-variable linear equation for x.

—If there is no solution to this new equation, there are no solutions to the system.

—If all real numbers are solutions to the new equation, there are infinitely many solutions; the two equations are **dependent.**

—Solve for y by plugging the x-value into the expression for y in terms of x.

—Check that the solution works by plugging it into the original equations.

Word Problems

The systematic way to solve word problems is to convert them to equations.

1. **Choose variables.** Choose wisely. Whatever you are asked to find usually merits a variable.

2. **Rewrite the statements** given in the problem as equations using your variables. Use common sense: more, fewer, sum, total, difference mean what you want them to mean. Common trigger words include:
   Of: Frequently means multiplication. **Ex:** "Half of the flowers are blue" means that if there are c flowers, then there are $\frac{1}{2}c$ blue flowers.
   Percent (%): Divide by 100. **Ex:** "12% of the flowers had withered" means that $\frac{12}{100}c$ flowers were withered.

3. **Solve the equation(s)** to find the desired quantity.

4. **Check that the answer makes sense.** If the answer is $3\frac{1}{4}$ girls in the park or −3 shoes in a closet, either you made a computational mistake or the problem has no solution.

RATE PROBLEMS

Rate problems often involve speed, distance, and time. These are often good variables candidates.

Equations to use:

$(\text{distance}) = (\text{speed}) \times (\text{time})$

$(\text{speed}) = \frac{\text{distance}}{\text{time}}$

$\text{time} = \frac{\text{distance}}{\text{speed}}$

Check that the units on distance and time correspond to the units on speed. Convert if necessary:

Time:
$1\,\text{min} = 60\,\text{s};\ 1\,\text{h} = 60\,\text{min} = 3{,}600\,\text{s}$

Distance: $1\,\text{ft} = 12\,\text{in};\ 1\,\text{yd} = 3\,\text{ft} = 36\,\text{in}$
$1\,\text{mi} = 1{,}760\,\text{yd} = 5{,}280\,\text{ft}$

Metric distance:
$1\,\text{m} = 100\,\text{cm};\ 1\,\text{km} = 1{,}000\,\text{m}$
$1\,\text{in} \approx 2.54\,\text{cm};\ 1\,\text{m} \approx 3.28\,\text{ft};\ 1\,\text{mi} \approx 1.61\,\text{km}$
$\text{Average speed} = \frac{\text{total distance}}{\text{total time}}$
Average speed is not the average of speeds used over equal distances, it's the average of speeds used over equal time intervals.

Ex: Supercar travels at $60\,\text{mi/h}$ for $30\,\text{min}$ and at $90\,\text{mi/h}$ for the rest of its 45-mile trip. How long does the trip take?

—The first part of the journey takes $30\,\text{min} \times \frac{1\,\text{h}}{60\,\text{min}} = \frac{1}{2}\,\text{h}$. During this time, Supercar travels $60\,\text{mi/h} \times \frac{1}{2}\,\text{h} = 30\,\text{mi}$.

—The second part of the trip is $45\,\text{mi} - 30\,\text{mi} = 15\,\text{mi}$ long. Supercar zips through this part in $\frac{15\,\text{mi}}{90\,\text{mi/h}} = \frac{1}{6}\,\text{h}$.

—The total trip takes $\frac{1}{2}\,\text{h} + \frac{1}{6}\,\text{h} = \frac{2}{3}\,\text{h}$, or $40\,\text{min}$.

What is Supercar's average

$\frac{\text{Total distance}}{\text{Total time}} = \frac{45\,\text{mi}}{\frac{2}{3}\,\text{h}} = 6$

This may seem low, bu traveled at $60\,\text{mi/h}$ and fourth of the journey wa

TASK PROBLEM

Ex: Sarah can paint a ho can do it in five. How lo together?

These problems are di Sarah paints a house in of $\frac{1}{4}$ house per day. Ju house per day. Workin complete one house:

$\frac{x}{5} + \frac{x}{4} = 1$.

Simplifying, we get $\frac{9x}{20}$ This makes sense: two S 2 days; two Justins can and a Justin need some

ADDING AND SUBTRACTING POLYNOMIALS

and $4y^5 - 16y^6$ are

in one variable:

ber of terms:
: **monomial**
: **binomial**
: **trinomial**

—Only like terms can be added or subtracted together into one term:
Ex: $3x^3y - 5xyx^2 = -2x^3y$.

MULTIPLYING POLYNOMIALS

The key is to multiply every term by every term, term by term. The number of terms in the (unsimplified) product is the product of the numbers of terms in the two polynomials.
—Multiplying a monomial by any other polynomial: Distribute and multiply each term of the polynomial by the monomial.

—Multiplying two binomials: Multiply each term of the first by each term of the second:
$(a + b)(c + d) = ac + ad + bc + bd$.

—MNEMONIC: **FOIL:** Multiply the two First terms, the two Outside terms, the two Inside terms, and the two Last terms.

—Common products:
$(a + b)^2 = a^2 + 2ab + b^2$
$(a - b)^2 = a^2 - 2ab + b^2$
$(a + b)(a - b) = a^2 - b^2$
—After multiplying, simplify by combining like terms.

$\begin{cases} x - 4y = 1 \\ 2x - 11 = 2y \end{cases}$

ing the first equation to solve for y in terms x gives $y = \frac{1}{4}(x - 1)$. Plugging in to ond equation gives $2x - 11 = 2\left(\frac{1}{4}(x - 1)\right)$. ving for x gives $x = 7$. Plugging in for y es $y = \frac{1}{4}(7 - 1) = \frac{3}{2}$. Check that $(7, \frac{3}{2})$ rks.

OLVING BY ADDING OR UBTRACTING QUATIONS: NO VARIABLES

xpress both equations in the same form. $ax + by = c$ works well.
ook for ways to add or subtract the equations to eliminate one of the variables.
·If the coefficients on a variable in the two equations are the same, subtract the equations.
·If the coefficients on a variable in the two equations differ by a sign, add the equations.
·If one of the coefficients on one of the variables (say, x) in one of the equations is 1, multiply that whole equation by the x-coefficient in the other equation; subtract the two equations.
·If no simple combination is obvious, simply pick a variable (say, x). Multiply the first equation by the x-coefficient of the second equation,

multiply the second equation by the x-coefficient of the first equation, and subtract the equations.
If all went well, the sum or difference equation is in one variable (and easy to solve if the original equations had been in $ax + by = c$ form). Solve it.
—If by eliminating one variable, the other is eliminated too, then there is no unique solution to the system. If there are no solutions to the sum (or difference) equation, there is no solution to the system. If all real numbers are solutions to the sum (or difference) equation, then the two original equations are dependent and express the same relationship between the variables; there are infinitely many solutions to the system.
—Plug the solved-for variable into one of the original equations to solve for the other variable.
Ex: $\begin{cases} x - 4y = 1 \\ 2x - 11 = 2y \end{cases}$
Rewrite to get $\begin{cases} x - 4y = 1 \\ 2x - 2y = 11 \end{cases}$.
The x-coefficient in the first equation is 1, so we multiply the first equation by 2 to get $2x - 8y = 2$, and subtract this equation from the original second equation to get:
$(2 - 2)x + (-2 - (-8))y = 11 - 2$ or $6y = 9$, which gives $y = \frac{3}{2}$, as before.

MORE THAN TWO VARIABLES

There is a decent chance that a system of linear equations has a unique solution only if there are as many equations as variables.
—If there are too many equations, then the conditions are likely to be too restrictive, resulting in no solutions. (This is only actually true if the equations are "independent"—each new equation provides new information about the relationship of the variables.)
—If there are too few equations, then there will be too few restrictions; if the equations are not contradictory, there will be infinitely many solutions.
—All of the above methods can, in theory, be used to solve systems of more than two linear equations. In practice, graphing only works in two dimensions. It's too hard to visualize planes in space.
—Substitution works fine for three variables; it becomes cumbersome with more variables.
—Adding or subtracting equations (or rather, arrays of coefficients called **matrices**) is the method that is used for large systems.

---

speed for the trip?

,5 mi/h.

it's right: Supercar had at 90 mi/h, but only one- at the faster speed.

S

se in four days, while Justin g will it take them working

guised rate problems. If days, she works at a rate tin works at a rate of $\frac{1}{5}$ for x days, they have to

1, or $x = \frac{20}{9} \approx 2.2$ days. rahs can do the house in lo it in 2.5 days; a Sarah ngth of time in between.

Exponential notation is shorthand for repeated multiplication:
$3 \cdot 3 = 3^2$ and $(-2y) \cdot (-2y) \cdot (-2y) = (-2y)^3$.
In the notation a^n, a is the base, and n is the exponent. The whole expression is "a to the nth power," or the "nth power of a," or, simply, "a to the n."

a^2 is "a squared;" a^3 is "a cubed."

$(-a)^n$ is not necessarily the same as $-(a^n)$.

Ex: $(-4)^2 = 16$, whereas $-(4^2) = -16$. Following the order of operation rules, $-a^n = -(a^n)$.

RULES OF EXPONENTS

Product of powers: $a^m a^n = a^{m+n}$
If the bases are the same, then to multiply, simply add their exponents. **Ex:** $2^3 \cdot 2^8 = 2^{11}$.

Quotient of powers: $\frac{a^m}{a^n} = a^{m-n}$
If the bases of two powers are the same, then to divide, subtract their exponents.

Exponentiation powers: $(a^m)^n = a^{mn}$
To raise a power to a power, multiply exponents.

Power of a product: $(ab)^n = a^n b^n$

Quotient of a product: $\left(\frac{a}{b}\right)^n = \frac{a^n}{b^n}$

Exponentiation distributes over multiplication and division, but not over addition or subtraction.
Ex: $(2xy)^2 = 4x^2y^2$, but
$(2 + x + y)^2 \neq 4 + x^2 + y^2$.

Zeroth power: $a^0 = 1$
To be consistent with all the other exponent rules, we set $a^0 = 1$ unless $a = 0$. The expression 0^0 is undefined.

Negative powers: $a^{-n} = \frac{1}{a^n}$

We define negative powers as reciprocals of positive powers. This works well with all other rules. **Ex:** $2^3 \cdot 2^{-3} = \frac{2^3}{2^3} = 1$.
Also, $2^3 \cdot 2^{-3} = 2^{3+(-3)} = 2^0 = 1$.

Fractional powers: $a^{\frac{1}{n}} = \sqrt[n]{a}$
This definition, too, works well with all other rules.

Completing the Square

The procedure called **completing the square** uses the reasoning from the example above to express an equation in the form $ax^2 + bx + c = 0$ as an equivalent equation that looks like $a(x + p)^2 = d$. We already know how to solve equations in the form $a(x + p)^2 = d$; completing the square will enable us to solve any quadratic equation.

If a is 1, the situation is simpler. That means we just need to get $x^2 + bx + c = 0$ to look like $(x + p)^2 = d$.

If you expand $(x + p)^2 = d$, it becomes

$$x^2 + 2px + p^2 = d$$

Compare that to $x^2 + bx + c = 0$. The x^2 terms already look the same. So far, so good. If the x terms are the same, it means that $2p = b$—in other words, that $p = \dfrac{b}{2}$. That's the key: once you've got p, you can figure out everything else.

To express $x^2 + bx + c = 0$ as $(x + p)^2 = d$.

EXAMPLE: Complete the square

Complete the square to express $x^2 + 6x + 7 = 0$ in the form $(x + p)^2 = d$.

SOLUTION

1. Move the constant c to the right-hand side of the equation. We do this for convenience, to remove the c from the left-hand side.

$$x^2 + 6x + 7 = 0$$
$$x^2 + 6x + 7 - 7 = 0 - 7$$
$$x^2 + 6x = -7$$

2. **Complete the square:** Let $p = \dfrac{b}{2}$. Compute p^2, or $\dfrac{b^2}{4}$, and add it to both sides of the equation. (You now have a perfect square on the left-hand side; by adding $\dfrac{b^2}{4}$, you've *completed* it.)

    Since $b = 6$, we must have $p = 3$ and $p^2 = 9$. Adding 9 to both sides,

$$x^2 + 6x + 9 = -7 + 9$$
$$x^2 + 6x + 9 = 2$$

3. Now the left-hand side looks like $x^2 + 2px + p^2$. Rewrite that as $(x + p)^2$, and you're done!

$$x^2 + 6x + 9 = 2$$
$$(x + 3)^2 = 2$$

So $x^2 + 6x + 7 = 0$ can be rewritten as $(x + 3)^2 = 2$.

Try the next example on your own before reading the solution.

EXAMPLE: Complete square on $x^2 + bx + c = 0$
Complete the square to re-express $x^2 + 10x - 2 = 0$ in the form $(x + p)^2 = d$.

SOLUTION
Rewrite the equation:

$$x^2 + 10x - 2 = 0$$

Move the constant term to the other side of the equation:

$$x^2 + 10x = 2$$

Complete the square: $b = 10$, so $p = 5$ and $p^2 = 25$. Add 25 to both sides:

$$x^2 + 10x + 25 = 2 + 25 = 27$$

Rewrite:

$$(x + 5)^2 = 27$$

EXAMPLE: Complete square on perfect square
Complete the square to express $x^2 - 18x - 9 = 0$ in the form $(x + p)^2 = d$.

SOLUTION
Rewrite the equation:

$$x^2 - 18x - 9 = 0$$

Move the constant term to the right side:

$$x^2 - 18x = 9$$

Complete the square. The linear coefficient (b) is -18, so $p = -9$ and $p^2 = 81$. Add to both sides:

$$x^2 - 18x + 81 = 9 + 81$$

Rewrite:

$$(x - 9)^2 = 90$$

So far, we've been working with getting $x^2 + bx + c = 0$ to look like $(x + p)^2 = d$. Now we'll put the a back into the picture. Read through this example to see how it works.

EXAMPLE: Complete square on $ax^2 + bx + c = 0$

Complete the square to express $-2x^2 + 6x = 0$ in the form $a(x + p)^2 = d$.

SOLUTION

Rewrite the equation. There's no constant term:

$$-2x^2 + 6x = 0$$

Factor out -2 from the left-hand side:

$$-2(x^2 - 3x) = 0$$

Complete the square: the x coefficient (inside the parentheses) is -3; half of that is $-\frac{3}{2}$; square to get $\frac{9}{4}$. To get $\frac{9}{4}$ inside the parentheses on the left-hand side, add -2 times $\frac{9}{4}$ to both sides:

$$-2\left(x^2 - 3x + \frac{9}{4}\right) = -2 \cdot \frac{9}{4}$$

Rewrite:

$$-2\left(x - \frac{3}{2}\right)^2 = -\frac{9}{2}$$

Solving by Completing the Square

We're ready to use completing the square to solve a general quadratic equation in the form $ax^2 + bx + c = 0$.

EXAMPLE: *Solve quadratic by completing the square*
Complete the square to find all the solutions of $x^2 + 2x - 2 = 0$.

SOLUTION
First, note that $x^2 + 2x - 2$ cannot be factored by conventional methods. No pair of integers whose product is –2 and whose sum is 2 exists. So we cannot solve this equation by factoring.

Rewrite the equation:

$$x^2 + 2x - 2 = 0$$

Move the constant term to the other side:

$$x^2 + 2x = 2$$

Complete the square. Since $b = 2$, we have $p = 1$ and $p^2 = 1$. Add 1 to both sides:

$$x^2 + 2x + 1 = 2 + 1$$

Rewrite:

$$(x + 1)^2 = 3$$

Take the square root of both sides:

$$x + 1 = \pm\sqrt{3}$$

Finish solving both $x + 1 = \sqrt{3}$ and $x + 1 = -\sqrt{3}$ individually:

$$x + 1 = \sqrt{3} \qquad \text{or} \qquad x + 1 = -\sqrt{3}$$
$$\downarrow \qquad\qquad\qquad \downarrow$$
$$x = -1 + \sqrt{3} \quad \text{or} \quad x = -1 - \sqrt{3}$$

So the two solutions of $x^2 + 2x - 2 = 0$ are $x = -1 + \sqrt{3}$ and $x = -1 - \sqrt{3}$. This is sometimes written as $x = -1 \pm \sqrt{3}$.

Although completing the square is not a popular method for solving quadratic equations, it is nevertheless an important technique. For one thing, it's what makes the quadratic formula possible. For another, it allows us to eyeball the shape of the graph of $y = ax^2 + bx + c$, much like finding the slope and the y-intercept allows us to eyeball the shape of a linear graph. So we will return to completing the square in the next chapter.

The Quadratic Formula

If you take a general quadratic equation $ax^2 + bx + c = 0$ and use the completing-the-square method of solution, keeping track of all the variables, you'll get the **quadratic formula**, an exact expression for the solutions to a general quadratic equation.

Quadratic Formula

If a, b, and c are real numbers with $a \neq 0$, then the solutions to $ax^2 + bx + c = 0$ are given by

$$x = \frac{-b + \sqrt{b^2 - 4ac}}{2a} \quad \text{and} \quad x = \frac{-b - \sqrt{b^2 - 4ac}}{2a}$$

The two solutions are often written as $x = \dfrac{-b \pm \sqrt{b^2 - 4ac}}{2a}$.

THE DISCRIMINANT

The **discriminant** of the quadratic $ax^2 + bx + c$ is the number under the radical in the quadratic formula. Its sign determines the number of real-number solutions of the equation $ax^2 + bx + c = 0$.

- **Positive discriminant:** If the discriminant is positive, then the equation has two distinct real-number solutions.

  That's because the expressions that you get from the quadratic formula, $\dfrac{-b + \sqrt{b^2 - 4ac}}{2a}$ and $\dfrac{-b - \sqrt{b^2 - 4ac}}{2a}$, evaluate to two unequal real numbers.

For example, the discriminant of $x^2 + 5x + 4$ is $(5)^2 - 4(1)(4) = 9$, which is positive; accordingly, $x^2 + 5x + 4$ has two solutions (namely, $x = -1$ and $x = -4$).

- **Negative discriminant:** If the discriminant is negative, then the equation has no real-number solutions.

  That's because if $b^2 - 4ac$ is negative, then you cannot evaluate $\sqrt{b^2 - 4ac}$ as a real number, so the expressions that you get from the quadratic formula don't have real-number values.

  For example, the discriminant of $x^2 + 3x + 4$ is $(3)^2 - 4(1)(4) = -7$, which is negative: $x^2 + 3x + 4$ has no real solutions.

- **Zero discriminant:** If the discriminant is zero, then the equation has only one solution.

  That's because if $b^2 - 4ac = 0$, then the quadratic-formula expressions $\dfrac{-b + \sqrt{b^2 - 4ac}}{2a}$ and $\dfrac{-b - \sqrt{b^2 - 4ac}}{2a}$ evaluate to the same number, namely, $\dfrac{-b}{2a}$.

  For example, the discriminant of $x^2 + 4x + 4$ is $(4)^2 - 4(1)(4) = 0$, so $x^2 + 4x + 4$ has exactly one solution (namely, $x = -2$).

All of these ideas may look complicated at first, but they're not so bad when you get the hang of how to apply them. The box below goes over the exact same properties of the discriminant again, this time in the language of polynomial roots rather than solutions to a quadratic equation.

KEY POINTS

The Sign of the Discriminant The polynomial $ax^2 + bx + c$ has two, one, or no real roots, depending on the sign of the discriminant $b^2 - 4ac$.

Sign of $b^2 - 4ac$	Number of real roots	Roots
positive	2	$-b \pm \sqrt{b^2 - 4ac}$
negative	0	none
0	1	$-\dfrac{-b}{2a}$

Now try to find the discriminants of these polynomials.

EXAMPLE: Find discriminant
Find the discriminant of each polynomial to determine how many roots it has.

a. $x^2 - 2x + 3$

b. $-\frac{1}{2}x^2 + 2x - 2$

c. $6x^2 - x - 35$

d. $x^2 + 1$

e. $x^2 - x - 1$

SOLUTION

a. $b^2 - 4ac = (-2)^2 - 4(1)(3) = -8$: the discriminant is negative, so $x^2 - 2x + 3$ has no real roots.

b. $b^2 - 4ac = 2^2 - 4\left(-\frac{1}{2}\right)(-2) = 0$: the discriminant is zero, so $-\frac{1}{2}x^2 + 2x - 2$ has one real root.

c. $b^2 - 4ac = (-1)^2 - 4(6)(-35) = 1 + 24 \cdot 35 > 0$: the discriminant is positive, so $6x^2 - x - 35$ has two real roots.

d. $b^2 - 4ac = 0^2 - 4(1)(1) = -4$: the discriminant is negative, so $x^2 + 1$ has no real roots.

e. $b^2 - 4ac = (-1)^2 - 4(1)(-1) = 5$: the discriminant is positive, so $x^2 - x - 1$ has two real roots.

USING THE QUADRATIC FORMULA

EXAMPLE: Quadratic formula

Use the quadratic formula to solve each of these quadratic equations.

a. $2x^2 + 6x = 0$

b. $x^2 + x - 1 = 0$

c. $x^2 + 5 = 0$

d. $-\frac{1}{2}x^2 + 2x - 2 = 0$

e. $6x^2 - x - 35 = 0$

SOLUTION

a.

$$x = \frac{-b \pm \sqrt{b^2 - 4ac}}{2a}$$

$$= \frac{-(6) \pm \sqrt{(6)^2 - 4(2)(0)}}{2(2)}$$

$$= \frac{-6 \pm 6}{4}$$

which simplifies to $x = \dfrac{-6 + 6}{4} = 0$ and $x = \dfrac{-6 - 6}{4} = -3$.

Since the two solutions are rational numbers, it's also possible to solve this equation by factoring. Indeed, $2x^2 + 6x = 2x(x + 3)$.

b.

$$x = \frac{-b \pm \sqrt{b^2 - 4ac}}{2a}$$

$$= \frac{-(1) \pm \sqrt{(1)^2 - 4(1)(-1)}}{2(1)}$$

$$= \frac{-1 \pm \sqrt{5}}{2}$$

So the two solutions are $x = \dfrac{-1 + \sqrt{5}}{2}$ and $x = \dfrac{-1 - \sqrt{5}}{2}$.

c.

$$x = \frac{-b \pm \sqrt{b^2 - 4ac}}{2a}$$

$$= \frac{-(0) \pm \sqrt{(0)^2 - 4(1)(5)}}{2(1)}$$

$$= \pm \frac{\sqrt{-20}}{2}$$

There is a negative number under a square root radical, so there are no solutions.

d.

$$x = \frac{-b \pm \sqrt{b^2 - 4ac}}{2a}$$

$$= \frac{-(2) \pm \sqrt{(2)^2 - 4\left(-\frac{1}{2}\right)(-2)}}{2\left(-\frac{1}{2}\right)}$$

$$= \frac{-2 \pm 0}{-1}$$

$$= 2$$

So there is only one solution, $x = 2$.

Since the solution is a rational number, this equation is possible to solve by factoring. Indeed,

$$-\frac{1}{2}x^2 + 2x - 2 = -\frac{1}{2}(x - 2)^2.$$

e.

$$x = \frac{-b \pm \sqrt{b^2 - 4ac}}{2a}$$

$$= \frac{-(-1) \pm \sqrt{(-1)^2 - 4(6)(-35)}}{2(6)}$$

$$= \frac{1 \pm \sqrt{841}}{12}$$

$$= \frac{1 \pm 29}{12}$$

which simplifies to $x = \frac{1 + 29}{12} = \frac{5}{2}$ and $x = \frac{1 - 29}{12} = -\frac{7}{3}$.

Since the solutions are rational numbers, it's possible to solve this equation by factoring. Indeed, $6x^2 - x - 35 = (2x - 5)(3x + 7)$.

Summary

Quadratic equations

A *quadratic equation* in one variable can always be expressed in the form $ax^2 + bx + c = 0$. A solution to the equation $ax^2 + bx + c = 0$ is also called a *root* of the polynomial $ax^2 + bx + c$.

Solving quadratic equations

There are four methods of solving a quadratic equation.

1. **Extracting roots:** Works on quadratic equations in the form $ax^2 = d$ and similar.

    - **Method:** Divide both sides by a, take square roots, and solve for x.

2. **Factoring:** Popular method but doesn't always work: many quadratics don't factor; many others are difficult to factor.

    - **Method:** Factor the polynomial $ax^2 + bx + c$. Then use the *Zero Product Property* to deduce that if $(x + p)(x + q) = 0$, then $x + p = 0$ or $x + q = 0$.

3. **Completing the square:** A tricky method, but it always works.

    - **Method:** Complete the square to express the polynomial as $a(x + p)^2 = d$. Then use the extracting-roots method.

4. **Quadratic formula:** Always works, but involves unpleasant calculation.

    - **Method:** The solutions to $ax^2 + bx + c = 0$ are given by

$$x = \frac{-b \pm \sqrt{b^2 - 4ac}}{2a}$$

The discriminant

There are three types of quadratic equations. Compute the *discriminant* $b^2 - 4ac$ of the quadratic $ax^2 + bx + c$ to find out which type of equation you're dealing with.

- If the discriminant is *positive*, there are two real solutions to $ax^2 + bx + c = 0$.

- If the discriminant is *negative*, there are no real solutions to $ax^2 + bx + c = 0$.

- If the discriminant is *zero*, there is exactly one solution to $ax^2 + bx + c = 0$.

Sample Test Questions

Answers to these questions begin on page 399.

1. Which of these are roots of the polynomial
   $x^2(2x - 1)(x + 3) + (1 - 2x)(3 + x)$?

   $$-3, \quad -1, \quad -\frac{1}{2}, \quad 0, \quad \frac{1}{2}, \quad 1, \quad 3$$

2. Which of these are quadratic equations?

   A. $3x - 7 = 8 - 4x$

   B. $\dfrac{4}{x + 1} - 5 = x^2$

   C. $\dfrac{x + 1}{4} - \sqrt[3]{5} = 2 - x^2$

   D. $2 - x^2 = \sqrt{x - 3} + 6$

   E. $x^2 - 7 = x^5 + x$

3. Solve each equation.

   A. $x^2 = 20$

   B. $x^2 + 45 = 0$

   C. $3x^2 = 84$

   D. $5x^2 - 27 = 0$

4. Find all solutions to each equation.

   A. $(x - 3)^2 = 4$

   B. $2(x + 5)^2 = 0$

   C. $3(x - 1)^2 = 15$

   D. $(2x + 3)^2 - 49 = 0$

5. Find the roots of each polynomial using the zero product property.

   A. $(x + 3)(x + 8)$

   B. $3z(z - 7)$

   C. $(2x - 9)(3x + 5)$

   D. $\left(x + \dfrac{3}{4}\right)\left(\dfrac{2}{3}x - \dfrac{5}{9}\right)$

6. Solve by factoring.

   A. $x^2 - 7x + 10 = 0$

   B. $y^2 - 18 = 7y$

   C. $2x^2 + 5x = 0$

   D. $4x^2 + 28x = -49$

   E. $3x^2 + 22x - 16 = 0$

   F. $(t - 2)(t - 3) = 30$

7. Fill in the missing term for each of these perfect-square polynomials.

   A. $x^2 + \underline{\quad} + 25$

   B. $z^2 - \underline{\quad} + 81$

   C. $x^2 + \underline{\quad} + \dfrac{49}{4}$

   D. $16x^2 + \underline{\quad} + 9$

8. Find the constant that must be added to each polynomial to make it a perfect square.

   A. $x^2 + 16x$

   B. $y^2 - 26y$

   C. $x^2 - 5x$

   D. $p^2 + \dfrac{1}{3}p$

9. Solve by completing the square.

   A. $y^2 - 10y = 24$

   B. $x^2 + 3x = -4$

   C. $x^2 - 2x - 5 = 0$

   D. $2x^2 - 10x - 28 = 0$

10. Find the discriminant of each polynomial and determine the number of its distinct real roots.

   A. $2x^2 - 5x - 3$

   B. $5x^2 - 6x + \dfrac{9}{5}$

   C. $x^2 - 4x + 13$

   D. $x^2 + 8x + 10$

11. Use the quadratic formula to find all solutions to each equation.

   A. $2x^2 - 5x - 3 = 0$

   B. $5x^2 - 6x + \dfrac{9}{5} = 0$

   C. $x^2 - 4x + 13 = 0$

   D. $x^2 + 8x + 10 = 0$

   Check that your answers are consistent with the answers for question 10.

12. Ella and Nina's unfortunately shaped dorm room is a rectangle that is 2.5 times longer than it is wide. If the floor area is 90 square feet, what are the dimensions of Ella and Nina's room?

   The area of a rectangle of length l and width w is given by $A = lw$.

13. Alex, who is 6 feet tall, tosses a rock straight up into the air at a speed of 46 feet per second. At time t after the throw, the rock is at a height of $-16t^2 + 46t + 6$ feet.

   A. How high is the rock 2 seconds after Alex lets go?

   B. How many seconds after the rock is thrown is it at a height of 25 feet?   *Hint:* There are two answers.

   C. If Alex lets the rock fall to the ground, how many seconds does the rock spend in the air?   *Hint:* The ground is at a height of 0 feet.

14. The longest side of a right triangle is one unit longer than twice the shortest side. The third side is seven units longer than the shortest side. What are the lengths of each of the three sides?   *Hint:* Use the Pythagorean theorem: if a right triangle has two shorter sides a and b and longest side c, then $a^2 + b^2 = c^2$.

Graphing Quadratic Equations

7

Overview

Graphs of linear equations are straight lines. Graphs of quadratic equations, on the other hand, have a curvy shape called a *parabola*. In this chapter we will explore parabolic shapes, both in relation to quadratic equations and on their own.

The parabola is a very special shape, both in nature and in science. Any object thrown up into the air follows a parabolic path. If you slice a cone parallel to its slant, the cross section will be a parabola. Parabolas also have special focusing properties: for example, a parabolic mirror reflects all incoming light into a single point. This makes parabolic shapes useful in optical instruments, including some contact lenses.

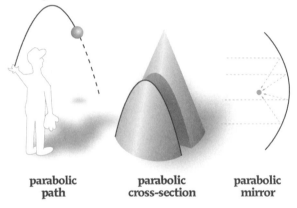

| parabolic path | parabolic cross-section | parabolic mirror |

Graphing Quadratic Equations

THE GRAPH OF $y = x^2$

Let's start by drawing the graph of $y = x^2$, the simplest quadratic equation.

The simplest way to start drawing a graph is to plot some points. The equation $y = x^2$ expresses y in terms of x, so pick a few x-values and plug in to find the corresponding y-values. You have some leeway in choosing x-values, but always try to include some positive values and some negative values. You should also usually include $x = 0$: it is easy to compute the y-value, and you immediately know the y-intercept of the graph.

x	y
0	
1	
2	
3	
4	
−1	
−2	
−3	

Try to fill in the table yourself before continuing.

x	y
0	0
1	1
2	4
3	9
4	16
−1	1
−2	4
−3	9

Note that opposite x-values, like -2 and 2, give the same y-value. That's because $a^2 = (-a)^2$ for all values of a. So $2^2 = (-2)^2$, $3^2 = (-3)^2$, and so on.

These eight points are all on the graph of $y = x^2$. Plotting them gives us a pretty good idea of the shape of the graph:

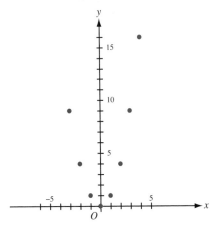

Connecting the points with a smooth curve reveals the graph of $y = x^2$:

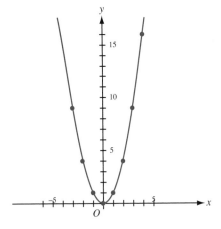

This curve is called a **parabola**. The graph of any quadratic equation $y = ax^2 + bx + c$ has the same basic shape.

Note that a parabola has one **axis of symmetry** that divides it into two mirror-image parts. The axis of symmetry passes through the **vertex**, which is the point where the parabola turns.

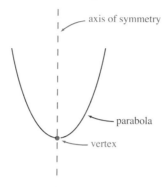

The parabola given by $y = x^2$, for example, has its vertex at $(0, 0)$ and its axis of symmetry along the y-axis. The vertex at $y = 0$ means that the minimum value of x^2 is 0. This symmetry happens because, as mentioned earlier, $a^2 = (-a)^2$ for every point a. So if the point $(2, 4)$ is on the graph, so is the point $(-2, 4)$; if $(3, 9)$ is there, so is $(-3, 9)$.

THE GRAPH OF $y = ax^2$: STRETCHING, COMPRESSING, FLIPPING

Now that we have an idea of the shape of the graph of $y = x^2$, we can investigate how the value of the coefficient a affects the shape of the graph of $y = ax^2$.

- If a has large absolute value, the parabola is narrower than $y = x^2$.

- If a has small absolute value, the parabola is wider than $y = x^2$.

- If a is negative, the parabola is flipped over the x-axis.

The Graph of $y = 2x^2$ and Similar

EXAMPLE: Graphing $y = 2x^2$

Graph $y = 2x^2$.

SOLUTION

First, we choose some x-values: some negative, some positive, and zero. We choose –3, –2, –1, 0, 1, 2, and 3.

Next, we make a table of values.

x	y
–3	18
–2	8
–1	2
0	0
1	2
2	8
3	18

Finally, we plot these points on the xy-plane and draw a parabolic curve through them.

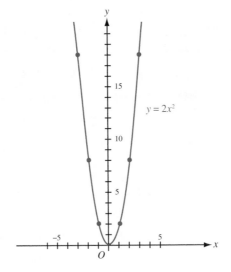

Like the graph of $y = x^2$, the graph of $y = 2x^2$ is a parabola. The vertex is still at (0, 0), and the axis of symmetry is still the y-axis. But the $y = 2x^2$ parabola is narrower than the parabola for $y = x^2$.

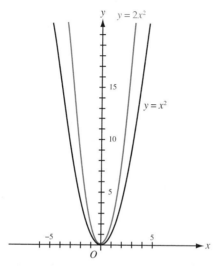

The $y = 2x^2$ parabola is what you get when you **stretch** the $y = x^2$ parabola **vertically** by a factor of 2. In other words, every point has moved vertically twice as far away from the x-axis. Given a particular value of x, the value of y that comes from $y = 2x^2$ is exactly twice the value that you get from $y = x^2$.

Look at the graphs of $y = x^2$, $y = 2x^2$, $y = 3x^2$, and $y = 4x^2$, all plotted on the same set of axes:

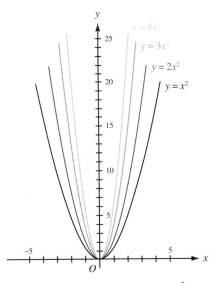

You can see that for $a > 1$, the graph of $y = ax^2$ is a parabola narrower than $y = x^2$; the greater a, the narrower the parabola.

The Graph of $y = -x^2$ and Similar

Work through this example yourself before looking at the answer.

EXAMPLE: Graphing $y = -x^2$

Graph the equation $y = -x^2$. How does this graph compare to the graph of $y = x^2$?

SOLUTION

First, choose some values of x and make a table of values:

x	y
−3	−9
−2	−4
−1	−1
0	0
1	−1
2	−4
3	−9

Again, $x = 1$ and $x = -1$ give the same y-value, as do $x = 2$ and $x = -2$ and so on. That's because $-(p)^2 = -(-p)^2$ for all values of p.

Now plot these points and connect them with a smooth curve.

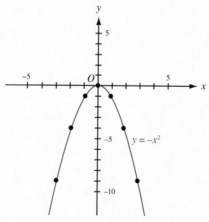

Compare this graph to the graph of $y = x^2$. The two are mirror images of each other. The axis of symmetry for the $y = -x^2$ graph is still the y-axis. The vertex is still at (0, 0), but this time the parabola opens down, so the vertex occurs at the highest point on the graph (the point with the greatest y-value), not the lowest.

We say that the graph of $y = -x^2$ is the **reflection** of the graph of $y = x^2$ **over the x-axis**.

EXAMPLE: Sketching $y = -ax^2$ from $y = ax^2$

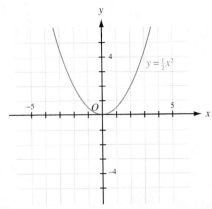

Use the graph of $y = \dfrac{1}{2}x^2$ above to sketch the graph of $y = -\dfrac{1}{2}x^2$ on the same set of axes.

SOLUTION

The graph of $y = -\dfrac{1}{2}x^2$ is the mirror image of the graph of $y = \dfrac{1}{2}x^2$ below the x-axis. Sketch the graph by choosing points and plotting points with the same x-coordinate and the opposite y-coordinate.

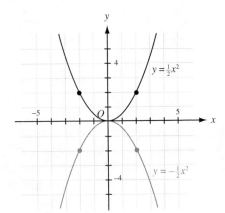

The Graph of $y = ax^2$

We can summarize the results of this section in the following table:

	$\lvert a \rvert > 1$	$\lvert a \rvert < 1$ small
$a > 0$	$a > 1$	$0 < a < 1$
	stretched vertically by a factor of a above the x-axis	compressed vertically by a factor of $\frac{1}{a}$ above the x-axis
$a < 0$	$a < -1$	$-1 < a < 0$
	stretched vertically by a factor of $\lvert a \rvert$ below the x-axis	compressed vertically by a factor of $\frac{1}{a}$ below the x-axis

If we again treat compressions as stretches by a factor less than 1, we can say that the graph of $y = ax^2$ is a vertical stretch of the graph of $y = x^2$ by a factor of $\lvert a \rvert$. If $a > 0$, the parabola opens up above the x-axis; if $a < 0$, the parabola opens down below the x-axis.

CHAPTER 7
GRAPHING QUAD. EQUATIONS

EXAMPLE: Classifying $y = ax^2$

For each equation, specify whether the graph opens up or down and whether the parabola is wider or narrower than $y = x^2$.

a. $y = -2x^2$

b. $y = -\dfrac{3}{2}x^2$

c. $y = 0.001x^2$

d. $y = -\dfrac{1}{0.001}x^2$

e. $y = \dfrac{3}{5}x^2$

SOLUTION

a. Since $a = -2$ is negative, the parabola opens *down*. Since $|a| = 2 > 1$, the graph is *narrower* than that of $y = x^2$.

b. Since $a = -\dfrac{3}{2}$ is negative, the parabola opens *down*. Since $|a| = \dfrac{3}{2} > 1$, the parabola is *narrower* than $y = x^2$.

c. Since $a = 0.001$ is positive, the parabola opens *up*. Since $|a| = 0.001 < 1$, the parabola is *wider* than $y = x^2$.

d. Since $a = -\dfrac{1}{0.001}$ is negative, the parabola opens *down*. Since $|a| = \dfrac{1}{0.001} = 1000 > 1$, the parabola is *narrower* than $y = x^2$.

e. Since $a = \dfrac{3}{5}$ is positive, the parabola opens *up*. Since $|a| = \dfrac{3}{5} < 1$, the parabola is *wider* than $y = x^2$.

EXAMPLE: Sketching $y = ax^2$

Sketch the graph of $y = -\dfrac{3}{2}x^2$.

SOLUTION

The equation is in $y = ax^2$ form, with $a = -\dfrac{3}{2}$. Since a is negative, the graph is a parabola that opens down. It is a vertical stretch of $y = -x^2$ by a factor of $\dfrac{3}{2}$.

Sketch the graph: a narrower parabola that opens down and has its vertex at (0, 0).

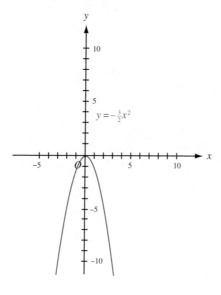

THE GRAPH OF $y = x^2 + h$: MOVING UP AND DOWN

Now we'll examine what happens when a constant is added to $y = x^2$. The parabola $y = x^2 + h$ has the same shape as $y = x^2$, but it is moved h units up or down, depending on the sign of h.

EXAMPLE: Graphing $y = x^2 + h$
Sketch the graph of $y = x^2 + 1$.

SOLUTION
First, make a table of values.

x	y
−3	10
−2	5
−1	2
0	1
1	2
2	5
3	10

Next, plot the points and draw a smooth curve through them to sketch the graph.

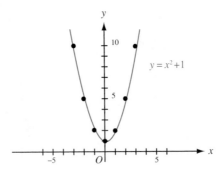

The graph is a parabola that opens up with its vertex at (0, 1), exactly 1 unit higher than the $y = x^2$ parabola. We say that the parabola $y = x^2 + 1$ is a **vertical shift**, or **vertical translation**, of the parabola $y = x^2$, one unit up.

Let's see another example of a parabola in $y = x^2 + h$ form.

EXAMPLE: Graphing $y = x^2 - h$

Sketch the graph of the equation $y = x^2 - 4$.

SOLUTION

First, make a table of values.

x	y
−3	5
−2	0
−1	−3
0	−4
1	−3
2	0
3	5

Next, plot the points and connect them with a smooth curve.

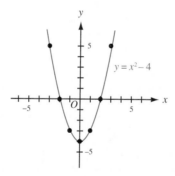

$$y = x^2 - 4$$

The parabola $y = x^2 - 4$ again has the same shape as $y = x^2$ but moved down 4 units, so that the vertex is at (0, −4). For each x-value, the value of y in $y = x^2 - 4$ is 4 less than in $y = x^2$.

Notice that the graph of $y = x^2 - 4$ crosses the x-axis twice, at $x = 2$ and $x = -2$. These are the roots of $x^2 - 4$. If you plug either into the polynomial, you get 0.

The Graph of $y = x^2 + h$

Let's summarize our findings thus far. The graph of $y = x^2$ is a parabola whose vertex is at $(0, 0)$. The graph of $y = x^2 + 1$ is the same parabola but with vertex at $(0, 1)$, one unit higher. The graph of $y = x^2 - 4$ is the same parabola again, with vertex at $(0, -4)$, four units lower than $y = x^2$.

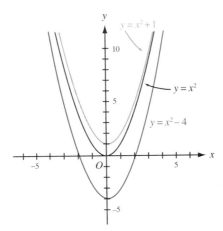

The graphs suggest that the graph of $y = x^2 + h$ is a vertical shift of the basic parabola by h units.

KEY POINTS

Graph of $y = x^2 + h$ The graph of $y = x^2 + h$ is a parabola that opens up with vertex at $(0, h)$.

It is a vertical shift of the $y = x^2$ parabola by h units—h units up if h is positive, $|h|$ units down if h is negative.

EXAMPLE: Sketching $y = x^2 + h$ from $y = x^2$

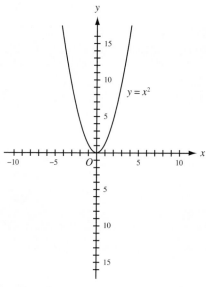

Use the graph of $y = x^2$ above to sketch a graph of $y = x^2 - 7$ on the same set of axes.

SOLUTION

The graph of $y = x^2 - 7$ is a parabola whose vertex is at $(0, -7)$. You can sketch it by shifting (or *translating*) the $y = x^2$ parabola down 7 units.

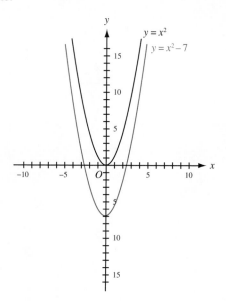

The Graph of $y = ax^2 + h$

We're ready to put the results of the last two sections together.

EXAMPLE: Graphing $y = ax^2 + h$

Sketch the graph of $y = 3x^2 - 10$.

SOLUTION

Make a table of values, plot the points, and connect them with a smooth curve.

x	y
−3	17
−2	2
−1	−7
0	−10
1	−7
2	2
3	17

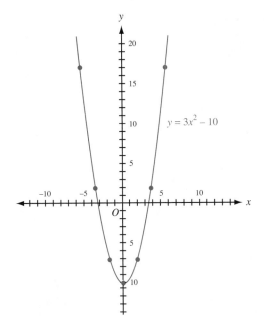

$y = 3x^2 - 10$

The graph of $y = 3x^2 - 10$ is again a parabola. It is narrower than the basic $y = x^2$ parabola: in fact, it has the same shape as the parabola $y = 3x^2$ except that the vertex is at $(0, -10)$. In short, it is a vertical shift of the graph of $y = 3x^2$ down 10 units. Of course, the graph of $y = 3x^2$ is itself a vertical stretch of the graph of $y = x^2$ by a factor of 3.

KEY POINTS

Graph of y = ax² + h The graph of $y = ax^2 + h$ is a parabola stretched and/or flipped like $y = ax^2$, and then shifted vertically h units. Its vertex is at $(0, h)$; it opens up if $a > 0$ and opens down if $a < 0$.

THE GRAPH OF $y = (x - k)^2$ MOVING LEFT AND RIGHT

We've looked at stretching, compressing, and flipping the parabola $y = x^2$, and we've looked at shifting it vertically. Now we look at shifting it horizontally.

EXAMPLE: Graphing $y = (x - k)^2$

Graph the equation $y = (x - 2)^2$.

SOLUTION

Make a table of values:

x	y
−3	25
−2	16
−1	9
0	4
1	1
2	0
3	1
4	4
5	9

Plot the points and connect them with a smooth parabolic curve.

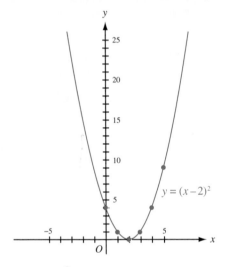

The graph of $y = (x - 2)^2$ is a parabola, the same shape as the $y = x^2$ parabola. But its vertex is at (0, 2), two units to the right of the origin. We say that the graph of $y = (x - 2)^2$ is a **horizontal shift**, or **horizontal translation**, of the graph of $y = x^2$, two units to the right.

Let's take a look at the graph of another equation in $y = (x - k)^2$ form. Try this example on your own—make a table of values, plot them, connect the dots—before looking at the solution.

EXAMPLE: Graphing $y = (x + k)^2$

Graph $y = (x + 3)^2$.

SOLUTION

Make a table of values:

x	y
−5	4
−4	1
−3	0
−2	1
−1	4
0	9
1	16
2	25

Plot these points and connect them with a smooth curve. The point (2, 25) is off the graph below.

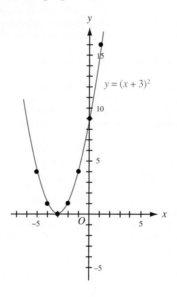

The graph of $y = (x + 3)^2$ is a parabola; its shape is the same as the shape of the parabola given by $y = x^2$, but its vertex is at $(-3, 0)$. The graph of $y = (x + 3)^2$ is a horizontal shift of the graph of $y = x^2$ three units to the left.

The Graph of y = (x – k)²

We've seen that the graph of $y = (x - 2)^2$ is a horizontal shift of the graph of $y = x^2$ two units to the right, vertex at $(2, 0)$, and that the graph of $y = (x + 3)^2$ is a horizontal shift of the graph of $y = x^2$ three units to the left, vertex at $(-3, 0)$.

We can make a guess that the graph of $y = (x - k)^2$ is, in general, a horizontal shift either left or right.

> *KEY POINTS*
>
> *Graph of y = (x ± k)²* The graph of $y = (x \pm k)^2$ is a parabola with the same shape as the graph of $y = x^2$.
>
> - If k is positive, then the graph of $y = (x - k)^2$ has its vertex at $(k, 0)$; it is a horizontal shift of the $y = x^2$ parabola k units to the *right*.
>
> - If k is positive, then the graph of $y = (x + k)^2$ has its vertex at $(-k, 0)$; it is a horizontal shift of the $y = x^2$ parabola k units to the *left*.

We often talk about shifts to the right (toward the positive x direction) as *positive* shifts and shifts to the left (toward the negative x direction) as *negative* shifts. So we can think of $y = (x - k)^2$ as a positive shift and $y = (x + k)^2$ as a negative shift. If we let k be either positive or negative, then $y = (x - k)^2$ is a positive shift if k is positive and a negative shift if k is negative.

> *KEY POINTS*
>
> *Graph of y = (x – k)², Redux* The graph of $y = (x - k)^2$ is a parabola with the same shape as the graph of $y = x^2$ and vertex at $(k, 0)$.
>
> It is a horizontal shift of the graph of $y = x^2$ by k units—k units to the right if k is positive, $|k|$ units to the left if k is negative.

Sometimes you'll have to fuss with minus signs to get the equation in the right form. For example, the equation $y = (x + 5)^2$ should be thought of as $y = (x - (-5))^2$, so that k is negative.

EXAMPLE: Sketch $y = (x - k)^2$ from $y = x^2$

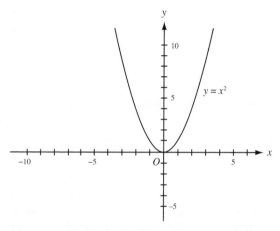

Use the graph of $y = x^2$ above to sketch the graph of $y = (x + 5)^2$ on the same set of axes.

SOLUTION
The graph of $y = (x + 5)^2$ is a horizontal shift of the graph of $y = x^2$, 5 units to the left.

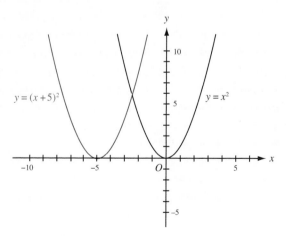

GRAPHING QUAD. EQUATIONS

CHAPTER 7

GRAPHING QUADRATIC EQUATIONS

THE GRAPH OF $y = a(x - k)^2 + h$: PUTTING IT TOGETHER

Flash review of what we've found so far:

- The graph of $y = x^2$ is a basic parabola with vertex at $(0, 0)$.
- The graph of $y = ax^2$ is a vertical stretch or compression of the basic parabola. If $a < 0$, the graph is also flipped over the x-axis.
- The graph of $y = x^2 + h$ is a basic parabola with vertex at $(0, h)$.
- The graph of $y = (x - k)^2$ is a basic parabola with vertex at $(k, 0)$.

These observations can be combined together, so that we can completely describe the graph of any equation in the form $y = a(x - k)^2 + h$.

KEY POINTS

The Graph of $y = a(x - k)^2 + h$ The graph of $y = a(x - k)^2 + h$ is a parabola.

- The vertex is at (k, h). The axis of symmetry is the vertical line $x = k$.
- If a is positive, the parabola opens up. If a is negative, the parabola opens down.
- If $|a| > 1$, the parabola is narrow, a vertical stretch of the $y = x^2$ parabola by a factor of a. If $|a| < 1$, the parabola is wide, a vertical compression of the $y = x^2$ parabola by a factor of $\dfrac{1}{|a|}$.

Again, rewrite an equation like $y = \dfrac{1}{2}(x + 3)^2 - 1$ as $y = \dfrac{1}{2}(x - (-3))^2 - 1$, so that k is negative.

Try this next example.

EXAMPLE: Reading off a, k, h

Each of these equations is in $y = a(x - k)^2 + h$ form. Give the values of a, k, and h for each.

a. $y = 2(x - 3)^2 + 5$

b. $y = -\frac{1}{4}(x + 6)^2 - 8$

c. $y = 0.92(x + 8.92)^2 + 4.32$

d. $y = -4(7 + x)^2 - 9$

e. $y = -2(8 - x)^2 - 2$

SOLUTION

a. $a = 2$, $k = 3$, $h = 5$

b. $a = -\frac{1}{4}$, $k = -6$, $h = -8$; note that k is negative

c. $a = 0.92$, $k = -8.92$, $h = 4.32$

d. $a = -4$, $k = -7$, $h = -9$

e. $a = -2$, $k = 8$, $h = -2$

Now try graphing.

EXAMPLE: Sketching $y = a(x - k)^2 + h$

Sketch the graph of $y = \frac{1}{2}(x + 3)^2 - 1$.

SOLUTION

The equation is in $y = a(x - k)^2 + h$ form, so we can read off its features. First, the vertex: $k = -3$ and $h = -1$, so the vertex is at $(-3, -1)$. Next, $a = \frac{1}{2}$, so the parabola opens up; it is wider than the $y = x^2$ parabola, a vertical compression by a factor of $\frac{1}{2}$.

Finally, sketch the graph: a wide parabola opening up with vertex at $(-3, -1)$.

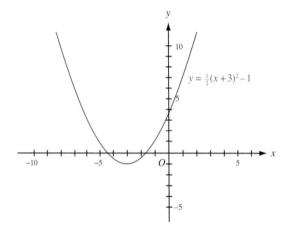

THE GRAPH OF $y = ax^2 + bx + c$: COMPLETING THE SQUARE

Sometimes the equation you need to graph isn't in the form $y = a(x - k)^2 + h$.

In order to graph a general quadratic equation in $y = ax^2 + bx + c$ form, first complete the square to express it in $y = a(x - k)^2 + h$ form and then sketch the parabola as we've been doing—vertex at (k, h), stretched and/or flipped depending on a.

EXAMPLE: Completing the square to sketch $y = ax^2 + bx + c$

Sketch the graph of $y = -x^2 + 3x - \dfrac{1}{4}$.

SOLUTION

First, complete the square to convert $-x^2 + 3x - \dfrac{1}{4}$ into

$a(x - k)^2 + h$ form. Keep the constant term separate and factor

out a –1 from $-x^2 + 3x$ to get

$$-1(x^2 - 3x) - \frac{1}{4}$$

We need to complete $x^2 - 3x$ as a perfect square. Half the x

coefficient is $-\dfrac{3}{2}$, so we're looking to get $\left(x - \dfrac{3}{2}\right)^2$, which is

$x^2 - 3x + \dfrac{9}{4}$. So we can add and subtract $\dfrac{9}{4}$ to complete the

square:

$$-\left(x^2 - 3x + \frac{9}{4} - \frac{9}{4}\right) - \frac{1}{4}$$

and then pull the $-\dfrac{9}{4}$ out of the parentheses, remembering to

pick up a distributed minus sign:

$$-\left(x^2 - 3x + \frac{9}{4}\right) + (-1)\left(-\frac{9}{4}\right) - \frac{1}{4}$$

Finally, simplify:

$$-\left(x - \frac{3}{2}\right)^2 + 2$$

So the equation can be written as $y = -\left(x - \dfrac{3}{2}\right)^2 + 2$.

Now we can read off the coordinates of the vertex and
determine whether the parabola is stretched or flipped:

- $k = \dfrac{3}{2}$ and $h = 2$, so the vertex is at $\left(\dfrac{3}{2}, 2\right)$
- $a = -1$, so the parabola has the same shape as $y = x^2$ but
  opens down

Finally, sketch the graph: a parabola with vertex at $\left(\frac{3}{2}, 2\right)$ opening down:

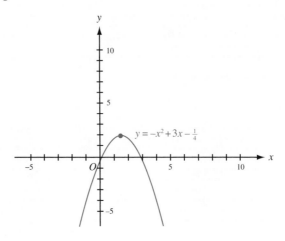

$$y = -x^2 + 3x - \tfrac{1}{4}$$

The Parabola and Its Properties

THE VERTEX

Finding the Vertex

Every parabola has a *vertex*, the point where the direction of the graph changes.

The vertex of a parabola given by $y = a(x - k)^2 + h$ is at (k, h), as we observed earlier.

The x-coordinate of the vertex of a parabola given by $y = ax^2 + bx + c$ is at $-\frac{b}{2a}$. (This happens to be what you get for k when you start completing the square.) To find the y-coordinate of the vertex, plug $x = -\frac{b}{2a}$ into the equation.

KEY POINTS

The Graph of $y = ax^2 + bx + c$ The graph of the equation $y = ax^2 + bx + c$ is a parabola whose vertex is at

$$\left(-\frac{b}{2a}, h\right)$$

where h is the value of $ax^2 + bx + c$ for $x = -\frac{b}{2a}$.

EXAMPLE: *Find vertex from equation*
For each equation, give the coordinates of the vertex of its parabola.

a. $y = 3(x - 4)^2 + 7$

b. $y = -\frac{3}{7}(x + 2)^2 + 5$

c. $y = x^2 + 6x - 5$

d. $y = 3x^2 - 4x$

e. $y = a(x - k)^2$

SOLUTION

a. Since k is 4 and h is 7, the vertex is at $(4, 7)$.

b. Since k is -2 and h is 5, the vertex is at $(-2, 5)$.

c. The x-coordinate of the vertex is $-\frac{6}{2 \cdot 1} = -3$. The y-coordinate is $(-3)^2 + 6(-3) - 5 = 9 - 18 - 5 = -14$. So the vertex is at $(-3, -14)$.

d. The x-coordinate of the vertex is $-\frac{-4}{2 \cdot 3} = \frac{2}{3}$. The y-coordinate is $3\left(\frac{2}{3}\right)^2 - 4\left(\frac{2}{3}\right) = 3\left(\frac{4}{9}\right) - \frac{8}{3} = \frac{4}{3} - \frac{8}{3} = -\frac{4}{3}$. So the vertex is at $\left(\frac{2}{3}, -\frac{4}{3}\right)$.

e. The vertex is at $(k, 0)$.

Maximum and Minimum Values

If a parabola opens up, the vertex is the least, or **minimum**, *y*-value on the graph; if a parabola opens down, then the vertex is the greatest, or **maximum**, *y*-value.

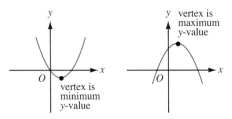

An up-opening parabola does not have a greatest *y*-value, nor does a down-opening parabola have a least *y*-value. The two branches of the parabola extend infinitely away from the vertex.

EXAMPLE: Find minimum of y-value of quadratic

What is the minimum *y*-value of the equation $y = 3(x - 4)^2 + 5$?

SOLUTION

We can answer this question in two ways, algebraically and geometrically. The geometric argument is simpler, but some instructors also teach the algebraic argument.

1. **Geometric solution:** The vertex of the parabola graph to $y = 3(x - 4)^2 + 5$ is at $(4, 5)$. The stretch factor 3 is positive, so the parabola opens up. This means that the minimum *y*-value of the equation is at the vertex, so the minimum *y*-value is 5.

2. **Algebraic solution:** We know that the smallest square of a real number is 0, since there are no negative squares. So no matter the value of *x*, the expression $(x - 4)^2$ will never be less than 0. Since $(x - 4)^2$ actually equals 0 for $x = 4$, we can safely say that the minimum value of $(x - 4)^2$ is 0. This means that the minimum value of $3(x - 4)^2$ is also 0 and that the minimum value of $3(x - 4)^2 + 5$ is $0 + 5 = 5$. So the minimum *y*-value of the equation $y = 3(x - 4)^2 + 5$ is 5.

This example suggests that you can read off the minimum (or maximum) *y*-value of an equation in $y = a(x - k)^2 + h$ form very easily: the vertex is at (k, h), so the minimum (or maximum) *y*-value is *h*.

KEY POINTS

Maximum or Minimum Value of a Quadratic

- If a is positive, then the graph of $y = a(x - k)^2 + h$ has a minimum y-value, at h. The parabola opens up and has no maximum value.

- If a is negative, then the graph of $y = a(x - k)^2 + h$ has a maximum y-value, at h. The parabola opens down and has no minimum value.

EXAMPLE: Finding max or min of a quadratic

For each equation, find the minimum or maximum value of y, as specified.

a. $y = (x - 2)^2 + 9$, minimum

b. $y = -\frac{1}{2}(x + 3)^2 - 4$, maximum

c. $y = 2(x - 3)^2 - 1$, maximum

d. $y = (x + 4)^2$, minimum

e. $y = 2x^2 - 12x + 7$, minimum

SOLUTION

a. The minimum y-value of $y = (x - 2)^2 + 9$ is 9.

b. The maximum y-value of $y = -\frac{1}{2}(x + 3)^2 - 4$ is –4.

c. The graph of $y = 2(x - 3)^2 - 1$ opens up, so there is no maximum y-value.

d. The minimum y-value of $y = (x + 4)^2$ is 0.

e. The minimum y-value of $y = 2x^2 - 12x + 7$ happens at the y-coordinate of the vertex, so you can plug in the x-coordinate of the vertex and solve for y. The x-coordinate of the vertex is $-\frac{b}{2a}$, or $-\frac{(-12)}{2(2)} = 3$; plugging $x = 3$ into $y = 2x^2 - 12x + 7$ gives $y = 2(3)^2 - 12(3) + 7 = -11$.

EXAMPLE: Determine and find max or min of quadratic

Each of these equations has either a maximum or a minimum
y-value. Say which it is and give that value.

a. $y = \frac{1}{2}(x-9)^2 + 8$

b. $y = -4x^2 - 7$

c. $y = -(x+2)^2$

d. $y = (x-2)^2 - 6$

e. $y = -3x^2 + 4x + 8$

SOLUTION

a. The stretch factor $a = \frac{1}{2}$ is positive, so there's a minimum
value at $h = 8$.

b. The stretch factor $a = -4$ is negative, so there's a maximum
value at $h = -7$.

c. The stretch factor $a = -1$ is negative, so there's a maximum
value at $h = 0$.

d. The stretch factor $a = 1$ is positive, so there's a minimum
value at $h = -6$.

e. The stretch factor $a = -3$ is negative, so there's a maximum

value at h, the y-coordinate of the vertex. The x-coordinate

of the vertex is $-\frac{b}{2a} = -\frac{(4)}{2(-3)} = \frac{2}{3}$, so the y-coordinate is

$y = -3\left(\frac{2}{3}\right)^2 + 4\left(\frac{2}{3}\right) + 8 = \frac{28}{3}$.

This means that $y = -3x^2 + 4x + 8$ is equivalent to

$y = -3\left(x - \frac{2}{3}\right)^2 + \frac{28}{3}$.

THE AXIS OF SYMMETRY

The *axis of symmetry* is an imaginary vertical line that goes through the vertex of a parabola. If you fold the coordinate plane in half along the axis of symmetry, the two halves of the parabola will match up exactly.

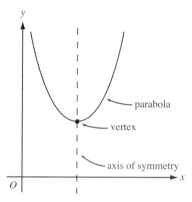

Since the axis of symmetry is a vertical line that goes through the vertex, its equation is $x = k$. This k is the same as in $y = a(x - k)^2 + h$, the x-coordinate of the vertex. From the previous section, we already know that $k = -\dfrac{b}{2a}$, so we can find the axis of symmetry directly from equations in $y = ax^2 + bx + c$ form or in $y = a(x - k)^2 + h$ form:

> **KEY POINTS**
>
> **Axis of Symmetry** The axis of symmetry for the parabola $y = ax^2 + bx + c$ is a vertical line with equation
> $$x = -\frac{b}{2a}$$
> Equivalently, the axis of symmetry for $y = a(x - k)^2 + h$ is given by $x = k$.

EXAMPLE: Using symmetry to sketch graph

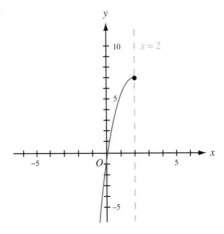

One-half of a parabola and its axis of symmetry are shown above. Complete the picture by drawing in the other half of the parabola.

SOLUTION

Find some points on the visible half of the parabola and flip them over the axis of symmetry to get some points on the other half of the parabola. For example, the parabola half shown goes through (1, 5) and (0, –1).

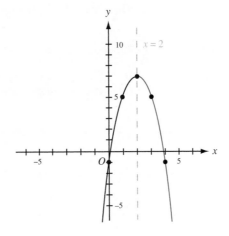

THE ROOTS

The *roots* of a polynomial are the x-values that make the polynomial evaluate to 0. So the roots of $ax^2 + bx + c$ are the solutions to $ax^2 + bx + c = 0$, or the x-intercepts of the graph of $y = ax^2 + bx + c$.

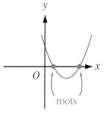

EXAMPLE: Roots, equation from graph

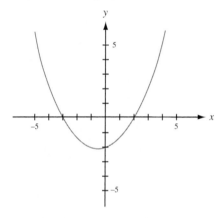

a. What are the roots of the polynomial whose graph is pictured above?

b. If the two roots are r and s, the equation of the parabola must be $y = a(x - r)(x - s)$. Find another point on the graph and use it to determine a and write down an equation for this parabola.

SOLUTION

a. The parabola has two x-intercepts, at $x = -3$ and at $x = 2$. These are the two roots.

b. Since the two roots are –3 and 2, the equation must have the form $y = a(x + 3)(x - 2)$. To compute a, we find a third point on the graph; the y-intercept $(0, -2)$ is a good pick. Now we can solve for a:

$$y = a(x + 3)(x - 2)$$
$$-2 = a(0 + 3)(0 - 2)$$
$$-2 = -6a$$
$$\frac{-2}{-6} = \frac{-6}{-6}a$$
$$\frac{1}{3} = a$$

So an equation for the parabola is $y = \frac{1}{3}(x + 3)(x - 2)$.

We can check this answer by expanding:
$y = \frac{1}{3}(x + 3)(x - 2) = \frac{1}{3}(x^2 + x - 6) = \frac{1}{3}x^2 + \frac{1}{3}x - 2$. The x-coordinate of the vertex should be at

$$-\frac{\frac{1}{3}}{2\left(\frac{1}{3}\right)} = -\frac{1}{2},$$

which seems just about right from the graph.

Counting the Roots

Recall that a quadratic polynomial $ax^2 + bx + c$ may have two, one, or zero real roots, depending on whether the discriminant $b^2 - 4ac$ is positive, 0, or negative, respectively.

These three scenarios play out with parabolas. A parabola may hit the x-axis two, one, or zero times.

A parabola with two real roots either has its vertex below the *x*-axis and opens up or it has its vertex above the *x*-axis and opens down.

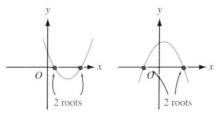

A parabola with one real root must have its vertex on the *x*-axis. It may open up or down.

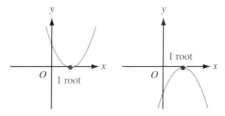

A parabola with no real roots must either be entirely above or entirely below the *x*-axis.

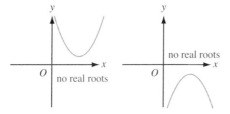

EXAMPLE: Classifying quadratic equations

Match each equation with its type of graph. (Each graph may be matched more than once or never.)

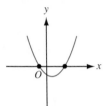
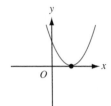
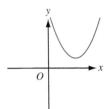

A. Opens up, two roots   **B**. Opens up, one root   **C**. Opens up, no roots

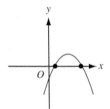
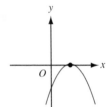
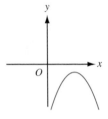

D. Opens down, two roots **E**. Opens down, one root **F**. Opens down, no roots

a. $y = -3x^2 + 4x - 7$

b. $y = -\dfrac{1}{2}x^2 - x + 5$

c. $y = (x + 2)(x - 4)$

d. $y = -2(x + 9)^2$

e. $y = 6(x - 3)^2 - 4$

SOLUTION

a. **F.** The discriminant of the polynomial is $b^2 - 4ac = 4^2 - 4(-3)(-7) = 16 - 84 = -68$. It is negative, so the polynomial has no real roots. Since $a = -3$ is negative, the parabola opens down.

b. **D.** The discriminant of the polynomial is $(-1)^2 - 4\left(-\dfrac{1}{2}\right)(5) = 1 + 10 = 11$. It is positive, so the polynomial has two roots. Moreover, since $a = -\dfrac{1}{2}$ is negative, the parabola opens down.

c. **A.** The polynomial is in factored form; it has two distinct roots (namely, $x = -2$ and $x = 4$). If multiplied out, the x^2 has a positive coefficient, so the parabola opens up.

d. **E.** The polynomial is in factored form; it has only one root (namely, $x = -9$). Since $a = -2$ is negative, the parabola opens down.

e. **A.** Since $a = 6$ is positive, this parabola opens up. Moreover, its vertex is at $(3, -4)$, below the x-axis. This means that both branches of the parabola cross the x-axis as they open up, so there are two roots.

Summary

Two forms of the quadratic equation

There are two common forms for the two-variable quadratic equation: $y = ax^2 + bx + c$ and $y = a(x - k)^2 + h$. (The a is the same.)

- Convert $y = ax^2 + bx + c$ to $y = a(x - k)^2 + h$ by *completing the square* (see page 290). It turns out that $k = -\dfrac{b}{2a}$ and $h = c - \dfrac{b^2}{4a^2}$. The a is the same.

- Convert $y = a(x - k)^2 + h$ to $y = ax^2 + bx + c$ by multiplying and simplifying. It turns out that $b = -2ak$ and $c = ak^2 + h$. The a is the same.

The parabola

The shape of the graph of a quadratic equation is called a *parabola*. Each parabola has a *vertex* and an *axis of symmetry*. A parabola opens either up or down.

Form of the equation	Vertex	Axis of symmetry	Direction
$y = ax^2 + bx + c$	$\left(-\dfrac{b}{2a}, c - \dfrac{b^2}{4a}\right)$	$x = -\dfrac{b}{2a}$	up if $a > 0$ down if $a < 0$
$y = a(x - k)^2 + h$	(k, h)	$x = k$	

Minimum and maximum values

A parabola that opens up has a minimum value, at h, but no maximum value. A parabola that opens down has a maximum value, at h, but no minimum value.

Shifting, stretching, and reflecting the parabola

The basic parabola $y = x^2$ has its vertex at the origin and opens up.

- **Vertical shift:** The graph of $y = x^2 + h$ is a vertical shift of the graph of $y = x^2$ by h units: up h units if h is positive, down $|h|$ units if h is negative.

- **Horizontal shift:** The graph of $y = (x - k)^2$ is a horizontal shift of the graph of $y = x^2$ by k units: right by k units if k is positive, left by $|k|$ units if k is negative.

- **Vertical stretch:** For positive a, the graph of $y = ax^2$ is a vertical stretch of the graph of $y = x^2$ by a factor of a.

- **Reflection over the *x*-axis:** The graph of $y = -x^2$ is a mirror image of the graph of $y = x^2$ over the *x*-axis.

Roots of a quadratic

The roots of $ax^2 + bx + c$ are the *x*-intercepts of the parabola $y = ax^2 + bx + c$. A parabola may touch the *x*-axis two, one, or zero times, depending on the sign of the discriminant $b^2 - 4ac$.

1. If $b^2 - 4ac > 0$, then the parabola crosses the *x*-axis twice. Equivalently, the quadratic has two real roots.

   It either has its vertex above the *x*-axis and opens down or it has its vertex below the *x*-axis and opens up.

   If the two *x*-intercepts are r and s, the equation of the parabola is $y = a(x - r)(x - s)$.

2. If $b^2 - 4ac = 0$, then the parabola touches the *x*-axis exactly once, at its vertex. Equivalently, the quadratic has exactly one root.

   It may open up or down, depending on the sign of a.

   The equation of the parabola is $y = a(x - k)^2$.

3. If $b^2 - 4ac < 0$, then the parabola does not cross the *x*-axis. Equivalently, the quadratic has no real roots.

   It either has its vertex above the *x*-axis and opens up or it has its vertex below the *x*-axis and opens down.

Sketching the graph of $y = ax^2 + bx + c$ or $y = a(x - k)^2 + h$

1. Determine whether the parabola **opens up or down** by looking at the sign of a.

2. **Plot the vertex** at (k, h) or $\left(-\dfrac{b}{2a}, c - \dfrac{b^2}{4a}\right)$.

3. (Recommended) **Plot the y-intercept** at c or $ak^2 + h$.

4. (Optional) **Plot the roots**, if any, especially if the quadratic is easy to factor.

5. **Sketch the parabola,** taking into account the stretch factor a.

Writing the equation of a parabola

1. If you can, **identify the roots** r and s. The equation will have the form $y = a(x - r)(x - s)$. All you still need to find is a.

   Otherwise, **identify the vertex** at (k, h). The equation will have the form $y = a(x - k)^2 + h$. All you still need to find is a.

2. To find a, **identify the y-intercept** at $(0, c)$ and plug its coordinates into your equation (either $y = a(x - r)(x - s)$ or $y = a(x - k)^2 + h$). Solve for a.

   If the y-intercept is hard to identify, use the coordinates of any other point on the graph.

Sample Test Questions

Answers to these questions begin on page 402.

1. Complete the square to rewrite each of these quadratic equations in $y = ax^2 + bx + c$ form as quadratic equations in $y = a(x - k)^2 + h$ form. Identify a, k, and h in each case.

   A. $y = x^2 - 4x + 9$

   B. $y = -x^2 - 7x$

   C. $y = 2x^2 + 8$

   D. $y = \frac{1}{3}x^2 + \frac{8}{3}x - 2$

   E. $y = -4x^2 + 8x - 4$

2. Find the coordinates of the vertex of the parabola given by each equation.

   A. $y = \frac{1}{2}(x + 7)^2 + 9$

   B. $y = 2x^2 + 4x - 6$

   C. $y = -\frac{2}{5}x^2 + 3x - 6$

   D. $y = -3x^2 + 5$

3. The graph of each of these equations is a parabola. What is the equation of its axis of symmetry?

   A. $y = x^2 - 12$

   B. $y = 3(x + 5)^2$

   C. $y = -15x^2 + 36x - 90$

4. The figure below shows the graph of $y = -2x^2 - 10x - 8$. Use it to find the roots of the polynomial $-2x^2 - 10x - 8$.

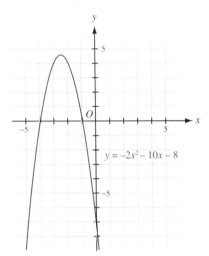

$y = -2x^2 - 10x - 8$

5. How many times does the graph of each equation cross or touch the x-axis?

   A. $y = (x + 2)(x - 7)$

   B. $y = -2x^2$

   C. $y = x^2 - 5x + 9$

   D. $y = 3x^2 - 8x - 2$

   E. $y = -5(x + 2)^2 + 4$

6. Find all the x- and y-intercepts of the graph of each equation.

   A. $y = (2x - 1)(x - 6)$

   B. $y = x^2 + 2x - 8$

   C. $y = 2x^2 - 7x + 9$

   D. $y = 9x^2 + 6x + 1$

   E. $y = -3(x + 4)^2 - 5$

7. Sketch the graph of each quadratic equation. Identify the vertex.

   A. $y = (x - 4)^2 - 2$

   B. $y = -3(x + 1)^2 + 5$

   C. $y = \frac{1}{2}(x - 2)^2 + 1$

   D. $y = -\left(x + \frac{7}{2}\right)^2$

8. Sketch the graph of each quadratic equation. Identify the vertex, the axis of symmetry, the y-intercept, and any roots.

   A. $y = -x^2 + 3$

   B. $y = 2x^2 - 10x$

   C. $y = -\frac{1}{3}x^2 + \frac{8}{3}x - \frac{7}{3}$

   D. $y = x^2 + 2x - 1$

9. Write down an equation for each parabola.

   A.

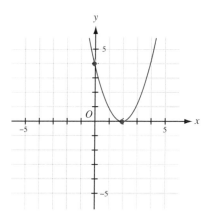

B.

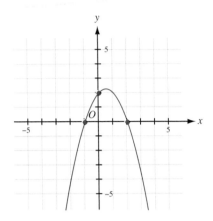

C.

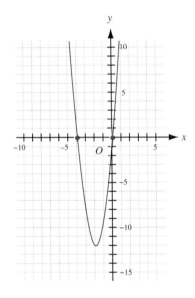

D.

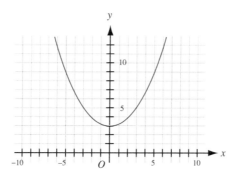

10.

   A. The vertex of a parabola is at (3, 2), and its y-intercept is –7. Give an equation for the parabola.

   B. The vertex of a parabola is at (–5, 0). It also passes through the point (–12, 4.9). Give an equation for the parabola.

11.

   A. A parabola has two x-intercepts, at –4 and 3, and a y-intercept at 2. Give an equation for the parabola.

   B. The only root of a polynomial is –2. Its graph passes through the point (1, –9). What is the polynomial?

12. The polynomial $a(x + 4)^2 – 7$ does not have any real roots. What is the sign of a?

13.

A. The graph of the cubic equation $y = x^3$ is pictured below. The graph of $y = x^3 + 2$ is a vertical shift of the graph of $y = x^3$, two units up. Sketch the graph of $y = x^3 + 2$ on the same set of axes below.

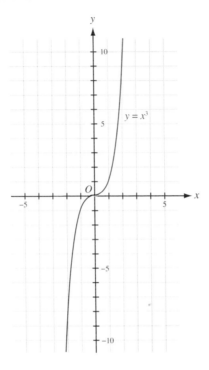

B. The graph of the cubic equation $y = x^3$ is pictured below. The graph of $y = -x^3$ is a mirror image of the graph of $y = x^3$, a reflection over the x-axis. Sketch the graph of $y = -x^3$ on the same set of axes below.

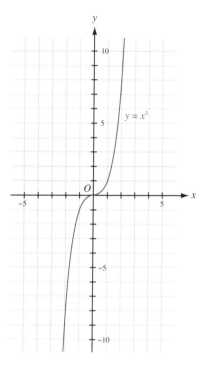

C. The graph of the cubic equation $y = x^3$ is pictured below. The graph of $y = (x + 4)^3$ is a horizontal shift of the graph of $y = x^3$ four units to the *left*. Sketch the graph of $y = (x + 4)^3$ on the same set of axes below.

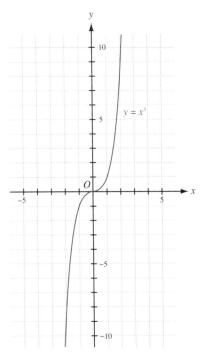

14. The graph of the equation $x = y^2 - 4y + 3$ is a parabola with a horizontal axis of symmetry. Make a table of values and sketch this parabola in the coordinate plane.

Inequalities

8

Overview

An equation contains an enormous amount of information: it says that two things are *exactly equal* to each other, as in $x + 3 = 7$ and $y^2 = 9$. Unfortunately, we don't always have the luxury of such precision. An *inequality* relates two things in a more vague way than an equation. Instead of knowing that $x + 3$ equals 7, we might know that $x + 3$ is *no less than* 7; instead of y^2 equaling 9, we might have that y^2 is *at most* 9.

The flip side of imprecision is leeway. The world is full of inexactitude where it seems that perfection should reign: tall buildings sway in the wind, rails expand on hot days, skin stretches without tearing, and bones bend without breaking. A little give is necessary, but a lot would be disaster. Inequalities are a language for expressing, analyzing, and at times controlling the imprecisions of our lives.

Inequalities and Solutions

INEQUALITY SIGNS

Suppose a and b are two real numbers. We can compare them in several ways.

< **less than**
$a < b$ if a is to the *left* of b along the number line.
Examples: $3 < 8$ and $-7 < -5$

> **greater than**
$a > b$ if a is to the *right* of b along the number line.
Examples: $9 > -4$ and $-1 > -100$

= **equal to**
$a = b$ if a and b identify the same point on the number line.
Examples: $-3 = -3$ and $2x = 2x$

≠ **not equal to**
   $a \neq b$ if a and b identify distinct points on the number line.
   Example: $2 + 2 \neq 5$

≤ **less than or equal to**
   Examples: $3 \leq 8$ and $3 \leq 3$

≥ **greater than or equal to**
   Examples: $-1 \geq -6$ and $-1 \geq -1$

The signs $<$, $>$, $\leq$, $\geq$, and $\neq$ are all **inequality signs**. The last one is $=$, the familiar **equal sign**.

KEY POINTS

Inequality Signs

1. Negative numbers with large absolute value are *less* than negative numbers with small absolute value. So $-100 < -1$ and $-5 > -7$. This is the opposite of what happens with positive numbers, where $100 > 1$ and $5 < 7$.

2. *Less than or equal to* means exactly what it says. It's true if either *less than* or *equal to* is true. Ditto for *greater than* or *equal to*. So 3 is *less than or equal to* 8 because 3 is *less than* 8. And -1 is *greater than or equal to* -1 because -1 is *equal to* -1.

3. What matters is the direction of the inequality sign, not which side the numbers are on. So $1 < 8$ and $8 > 1$ are the same statement, mathematically speaking. That being said, the two statements are read differently. The first is "one is less than eight," and the second one is "eight is greater than one."

The smaller side of an inequality sign faces the smaller number; the larger side faces the larger number. Also, if you see the sign as the gaping mouth of a hungry creature, the creature wants to eat the bigger number.

INTRODUCING INEQUALITIES

An **inequality** is a statement that says that two things are not equal. For example,

$$-2 < 7 \quad \text{and} \quad x \geq 8 \quad \text{and} \quad 2 < 2x + 1 \leq 10$$

are all inequalities. The last one is a **compound inequality**, two inequalities in one: namely, $2 < 2x + 1$ and $2x + 1 \leq 10$.

An inequality looks similar to an equation. Both are comparisons of two expressions. An *equation* says that two expressions are *equal* and connects them with an *equal* sign; an *inequality* says how two expressions are *unequal* and connects them with an *inequality* sign:

equation	inequality
$3x^2 - 1 = 7$	$3x^2 - 1 < 7$

TRUE AND FALSE INEQUALITIES

Like an equation, an inequality may be true or false.

For example, $2 > 1$ is a true inequality, as is $2 \geq 2$. But $2 < 1$ is a false inequality because 2 is not, in fact, less than 1.

Most inequalities that involve variables are true for some values of the variable and false for others. For example, $2x + 1 < 10$ is true for $x = 1$ because $2(1) + 1 < 10$ is a true statement. But the same inequality is false for $x = 8$ because $2(8) + 1 < 10$ is a false statement.

SOLUTIONS TO INEQUALITIES

A **solution** to an inequality in one variable is a value for the variable that makes the inequality into a true statement. For example, $z = 0$ is a solution to the inequality $z < 7$ because $0 < 7$ is a true statement.

We know that, most of the time, a linear equation in one variable has exactly one solution. In contrast, most inequalities have many, many solutions.

The inequality $z < 7$, for example, has $z = 0$ as a solution. But $z = 1$ is also a solution, since $1 < 7$. So are $z = -2$ and $z = \dfrac{2}{3}$ and $z = -1.8$. In fact, any real number less than 7 is a solution to $z < 7$. This makes it easy: you can look at a simple inequality like $z < 7$ and know right away what its solutions are.

EXAMPLE: Check solutions to inequality
Determine whether each number is a solution to the inequality $3y - 4 \le 8$.

a. 0

b. 5

c. $-\dfrac{1}{3}$

d. 4

SOLUTION

a. Yes, 0 is a solution: $3(0) - 4 = -4$, and $-4 \le 8$ is a true statement.

b. No, 5 is not a solution: $3(5) - 4 = 11$, and $11 \le 8$ is a false statement.

c. Yes, $-\dfrac{1}{3}$ is a solution: $3\left(-\dfrac{1}{3}\right) - 4 = -5$, and $-5 \le 8$ is a true statement.

d. Yes, 4 is a solution: $3(4) - 4 = 8$, and $8 \le 8$ is a true statement.

Solutions to most inequalities are continuous pieces of the real number line called *intervals* or *rays*. There are a couple of different way to describe these pieces; for this we need to look at some notation.

Sets and Intervals: Notation

SET NOTATION

A **set** is simply a collection of numbers, say, all the solutions to a particular equation or inequality. Sets are enclosed in curly braces {}: the set that contains the numbers 2, 3, and 7 is written

$$\{2, 3, 7\}$$

For example, you can say that the set of all solutions to the equation $(x + 1)(x - 2) = 0$ is $\{-1, 2\}$.

Roster notation is writing a set by listing all of its elements, as we just did for the set of $\{-1, 2\}$.

Set-builder notation is another way of writing down a set, by describing rather than listing its elements. For example, the set of real numbers less than 2 can be written as

$$\{x : x \text{ is a real number and } x < 2\}$$

Most of the time it's understood that x is a real number unless otherwise specified, so people just write

$$\{x : x < 2\}$$

This last is read as "the set of all x such that x is less than 2." The colon after the first x is shorthand for *such that*. Some writers use a vertical line instead of a colon: $\{x \mid x < 2\}$.

TRICKY POINTS

Dummy Variables The x in $\{x : x < 2\}$ is what's known as a *dummy variable*: its name (in this case, x) is not important. That is, you can write a different letter instead of x (say, z or a) without changing the thing you're talking about. So

$$\{a : a < 2\} = \text{the set of all } a \text{ such that } a < 2$$

is the exact same set of numbers as $\{x : x < 2\}$.

The set without any members is called the **empty set**. It is denoted as $\emptyset$ or as {} (but never as $\{\emptyset\}$).

Set Notation and Solutions to Inequalities

Set notation is one way of writing down the solutions to an inequality. The set of solutions to the inequality $x \le 3$ is, quite simply, the set $\{x : x \le 3\}$.

The difference between the two is just a matter of proper grammar: $x \le 3$ is a statement about the variable x, and $\{x : x \le 3\}$ is a collection of real numbers.

INTERVAL NOTATION

Open and Closed Intervals

An **interval** is a continuous segment of the number line. The interval

is all the numbers between *and including* 3 and 5, the same thing as $\{x : 3 \le x \le 5\}$. This interval may be written in interval notation as:

$$[3, 5]$$

where the square brackets mean that both 3 and 5, the two **endpoints** of the interval, are included. An interval like [3, 5], which includes its endpoints, is said to be **closed**.

An interval that does not include its endpoints is called **open**. For example, the open interval

is all the numbers between –2 and 1, where the open circles at –2 and 1 show that –2 and 1 are not included in the interval. This is the same set as $\{x : -2 < x < 1\}$.

Using interval notation, this set is written as

$$(-2, 1)$$

where the parentheses mean that neither endpoint is included in the interval.

> ### TRICKY POINTS
>
> *(a, b): Open Interval or Ordered Pair?* In the context of a coordinate plane, the notation (a, b) denotes the ordered pair whose first coordinate is a and whose second coordinate is b. To avoid confusion, math writers often clarify by writing "the open interval (a, b)" even when it's possible to guess which one is meant from context.

Intervals that include one endpoint but not the other are called **half-open** or, sometimes, **half-closed**. For example, the half-open interval

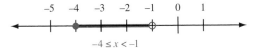

$$-4 \leq x < -1$$

is all the numbers between –4 and –1, including –4 (closed circle) but excluding –1 (open circle). This is the same set as $\{x : -4 \leq x < -1\}$.

In interval notation, this set is written as

$$[-4, 1)$$

Note the square bracket around the included endpoint and the parenthesis around the excluded endpoint.

Bounded and Unbounded Intervals

All the intervals that we've looked at so far have been **bounded**: they have two endpoints and a finite *length*.

Unbouded intervals do not have a set length. For example, the unbounded interval

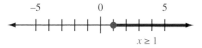

consists of all the numbers greater than or equal to 1, the same set as $\{x : x \geq 1\}$. This interval includes its only endpoint, so it is closed.

In interval notation, this interval is written as

$$[1, \infty)$$

where ∞ is a symbol for **infinity**.

> ### TRICKY POINTS
>
> *Infinity* Infinity is not a number; it's a notational aid that means that you include all the numbers to the right of 1, no matter how big. You can think of ∞ as a symbol representing something bigger than any real number.

Infinity itself is not included in the interval (it can't be—it's not a number), so it always appears with a parenthesis, never a square bracket. Nonetheless, $[1, \infty)$ is a closed set.

Intervals may also be unbounded to the left. For example, the unbounded interval

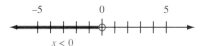

is all the numbers strictly less than 0; the set $\{x : x < 0\}$. In interval notation, this becomes

$$(-\infty, 0)$$

where $-\infty$ is **negative infinity**, representing something less than any real number. This interval does not include its endpoint, so it is open.

Unbounded intervals like the two we've just looked at are also called **rays**.

Interval Notation and Solutions to Inequalities

Interval notation is a second way of writing down solutions to inequalities.

For example, the set of solutions to the inequality $3 < x < 5$ is the open interval $(3, 5)$.

The set of solutions to the inequality $x \geq -7$ is the closed ray $[-7, \infty)$.

Graphing Inequalities

Three ways of representing the solutions of an inequality are:

1. **Set notation:** The set of solutions to $x > 8$ is $\{x : x > 8\}$.

2. **Interval notation:** The set of solutions to $4 < x \leq 6$ is $(4, 6]$.

3. **Graph:** The set of solutions to $x > 7$ is represented as the set of points on the number line that make the inequality into a true statement. In other words, all the numbers less than, or to the *left* of, 7.

$x < 7$

Observe the open circle at $x = 7$: it means that 7 is not included in the set of solutions. And indeed, $(7) < 7$ is not a true statement, so $x = 7$ is not a solution to the inequality.

GRAPHS OF BASIC INEQUALITIES

Try these examples on your own before looking at the solutions.

EXAMPLE: Graph open ray right
Graph the inequality $x > -1$.

SOLUTION
The graph is all the points greater than, or to the *right* of, -1.

The open circle at -1 shows that -1 is not included; the solution set is an open ray.

EXAMPLE: Graph closed ray right
Graph the inequality $x \geq 2$.

SOLUTION
The solutions are all values of x greater than or equal to 2—that is, to the right of and including 2.

A filled-in circle at 2 shows that 2 is a point on the graph.

EXAMPLE: Graph closed ray left
Graph the inequality $x \leq -\dfrac{3}{2}$ on the number line.

SOLUTION
The graph includes all the points less than or equal to $-\dfrac{3}{2}$, that is, to the left of and including the point $-\dfrac{3}{2}$ on the number line. The closed circle at $-\dfrac{3}{2}$ shows that the endpoint is included.

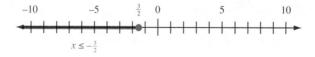

EXAMPLE: Graph x ≠ a

Graph the inequality $x \neq 1$.

SOLUTION

The graph includes every point on the number line except 1. Use an open circle at that point to indicate that it is not included in the graph.

EXAMPLE: Graph open interval

Graph the solutions to $2 < x < 8$.

SOLUTION

The inequality $2 < x < 8$ is a compound inequality; it is two inequalities put together:

$2 < x < 8$ means that $2 < x$ and, at the same time, $x < 8$.

Now, $2 < x$ is true for values of x to the right of 2—

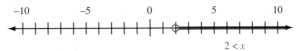

—and $x < 8$ is true for values of x to the left of 8—

That means that we're looking for values of x *both* to the right of 2 *and* to the left of 8. In other words, the compound inequality is true for values of x *between* 2 and 8:

EXAMPLE: Graph closed interval

Graph the interval $-3 \leq x \leq 4$.

SOLUTION

The values of x that satisfy $-3 \leq x \leq 4$ have to be greater than or equal to -3 *and* less than or equal to 4. These are the points on the number line between -3 and 4, inclusive.

EXAMPLE: Graph half-open interval

Graph the values of x that satisfy the compound inequality $-9 \leq x < -2.5$.

SOLUTION

The values of x that satisfy $-9 \leq x < -2.5$ are those between -9 and -2.5. The endpoint -9 is included but -2.5 is not.

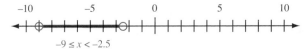

Solving Inequalities in One Variable

ALGEBRAIC METHODS

Solving inequalities is very similar to solving equations. The solutions will be intervals, or rays, rather than individual numbers.

The method of solution is the same: we do certain things to both sides of the inequality to get an **equivalent** but less complicated inequality, which has the same solutions but is easier to solve. We can either add the same number to both sides or multiply both sides by the same number. Adding works exactly the same way as with equations; multiplying is a little trickier—you have to watch out for negative signs.

Since adding is simpler, let's start there. Just like with equations, it's okay to add the same number to both sides of an inequality— adding the same number to both sides won't change an inequality's solutions.

KEY POINTS

Additive Property for Inequalities For any real numbers a, b, and c:

$$a < b \quad \text{is equivalent to} \quad a + c < b + c$$
$$a > b \quad \text{is equivalent to} \quad a + c > b + c$$
$$a \leq b \quad \text{is equivalent to} \quad a + c \leq b + c$$
$$a \geq b \quad \text{is equivalent to} \quad a + c \geq b + c$$

In other words, it's okay to add the same number to both sides of an inequality.

Similarly, it's okay to *subtract* the same number from both sides of an inequality.

EXAMPLE: Solve by adding, closed ray
Solve $x + 3 \geq 8$.

SOLUTION
Use the additive property for inequalities to add –3 to both sides. This will clear out the 3 on the right side and isolate the x:

$$x + 3 \geq 8$$
$$x + 3 - 3 \geq 8 - 3$$
$$x \geq 5$$

So the inequality $x + 3 \geq 8$ is equivalent to the inequality $x \geq 5$. Therefore $x + 3 \geq 8$ has the same solutions as $x \geq 5$, namely, $\{x : x \geq 5\}$:

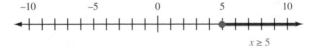

$$x \geq 5$$

Here's one more example that can be solved using only addition. This one's a compound inequality, but don't let that deter you— work on each inequality separately. Try to do this example on your own before reading the solution.

EXAMPLE: Solve by adding, open interval

Solve $-3 < x - 2 < 10$. Write down the solution set and graph it on a number line.

SOLUTION

This is a two-in-one inequality; work on each side separately:

$$-3 < x - 2 \qquad \text{and} \qquad x - 2 < 10$$

To isolate the x, add 2 to both sides—of each inequality:

$$-3 < x - 2 \qquad \text{and} \qquad x - 2 < 10$$
$$-3 + 2 < x - 2 + 2 \qquad \text{and} \qquad x - 2 + 2 < 10 + 2$$
$$-1 < x \qquad \text{and} \qquad x < 12$$

So the values of x that satisfy $-3 < x - 2 < 10$ are exactly the values that satisfy both $-1 < x$ and $x < 12$. In other words, the solutions are all the numbers in the interval $(-1, 12)$:

$$-1 < x < 12$$

Spot-check the answer by picking a point in the solution region and making sure it works in the inequality. The easiest number in the solution region is 0. Plug in: $-3 < 0 - 2$ is a true statement, and $0 - 2 < 10$ is a true statement. So 0 is definitely a solution, and our solution region looks pretty good.

In set notation, this is written as $\{x : -1 < x < 12\}$.

On to the second method of solving inequalities: multiplying both sides by the same number. If you multiply both sides of an inequality by a *positive* number, the inequality remains unchanged. That is, you get an equivalent inequality, which has the same solutions.

But if you multiply both sides of an inequality by a *negative* number, you *must flip the sign* of the inequality. Otherwise, the inequality you'll get won't be equivalent to the one with which you started.

KEY POINTS

Multiplicative Property for Inequalities

Part I

For any real numbers a, b, and *positive* real number c:

$$a < b \quad \text{is equivalent to} \quad ac < bc$$
$$a > b \quad \text{is equivalent to} \quad ac > bc$$

Similar statements are true for ≤ and ≥.

In other words, it's okay to multiply both sides of an inequality by a *positive* number.

It is also okay to *divide* both sides of an inequality by the same positive number.

Part II

For any real numbers a, b, and *negative* real number c:

$$a < b \quad \text{is equivalent to} \quad ac > bc$$
$$a > b \quad \text{is equivalent to} \quad ac < bc$$

Similar statements are true for ≤ and ≥.

In other words, if you multiply both sides of an inequality by a *negative* number, you must flip the inequality sign.

You can also *divide* both sides of an inequality by the same negative number if you also flip the inequality sign.

Do the first example on your own before reading the solution.

EXAMPLE: Solve by positive multiplying
Solve $3x \le 18$.

SOLUTION
Divide both sides of the inequality by 3 in order to isolate the x on the left side:

$$3x \le 18$$

$$\frac{3x}{3} \le \frac{18}{3}$$

$$x \le 6$$

So the solution set to $3x \le 18$ is $\{x : x \le 6\}$.

To spot-check the solution, you can pick a number from the solution set and plug it in. You should get a true statement. For example, 0 is in the set $\{x : x \le 6\}$. Plugging 0 into $3x \le 18$ gives $3(0) \le 18$, which is a true statement, a good sign for this solution.

Of course, you won't be able to catch all mistakes by this method, but it's a good start—and it will help tremendously if you've forgotten to flip the inequality sign when multiplying or dividing by a negative number.

EXAMPLE: Solve by negative multiplying
For what values of x is $-5x > 70$ true?

SOLUTION
Divide both sides of the inequality by -5 in order to isolate the x on the left side. Remember to flip the inequality sign.

$$-5x > 70$$

$$\frac{-5x}{-5} < \frac{70}{-5}$$

$$x < -14$$

So the solution set is $\{x : x < -14\}$.

Pick a value and plug it in to spot-check the solution. It's a good idea to do this whenever you've multiplied or divided by a negative number. Although 0 is *not* in the solution set, you can still use it as a checkpoint: when you plug it in, you should get a *false* statement. Fortunately, $-5(0) < -14$ is false.

CHAPTER 8
INEQUALITIES

Just as with equations, we can use the multiplication property and the addition property together to solve inequalities. Try both of these examples on your own before reading the solution.

EXAMPLE: *Solve by adding and positive multiplying*

Solve $4x + 10 \geq 3$.

SOLUTION

Rewrite the inequality:

$$4x + 10 \geq 3$$

Subtract 10 from both sides (or add –10) to isolate the x term on the left:

$$4x + 10 - 10 \geq 3 - 10$$
$$4x \geq -7$$

Divide both sides by 4 to isolate x:

$$\frac{4x}{4} \geq \frac{-7}{4}$$
$$x \geq -\frac{7}{4}$$

So the solution set is $\{x : x \geq -\frac{7}{4}\}$.

Spot-check the solution by plugging in a particular value. In this case 0 is in the solution set, so plugging it in should give a true statement. Indeed, $4(0) + 10 \geq 3$ is true.

EXAMPLE: Solve by adding and negative multiplying

Solve the compound inequality: $-7 < 9 - x < 9$.

SOLUTION

Rewrite the compound inequality as a combination of two simple inequalities:

$$-7 < 9 - x \qquad \text{and} \qquad 9 - x < 9$$

For both inequalities, subtract 9 from both sides to isolate the x term:

$$-7 - 9 < 9 - x - 9 \qquad \text{and} \qquad 9 - x - 9 < 9 - 9$$
$$-16 < -x \qquad\qquad \text{and} \qquad\qquad -x < 0$$

To get x instead of $-x$, multiply both sides of both inequalities by -1, flipping the inequality signs:

$$-1(-16) < -1(-x) \qquad \text{and} \qquad -1(-x) < -1(0)$$
$$16 > x \qquad\qquad \text{and} \qquad\qquad x > 0$$

So x must satisfy both $16 > x$ and $x > 0$. We can rewrite this as $0 < x < 16$, so the solution set is $\{x : 0 < x < 16\}$.

Spot-check by plugging in a particular value from the solution region. Any number will do; let's use 1. Plugging 1 into $-7 < 9 - x$ gives $-7 < 9 - 1$, a true statement. Plugging 1 into $9 - x < 9$ gives $9 - 1 < 9$, also a true statement. So 1 is a solution to the inequality, and the solution set $\{x : 0 < x < 16\}$ looks pretty good.

USING GRAPHS

Although it's good to know how to manipulate inequalities algebraically as we've been doing, you can solve most inequalities without worrying about which way the inequality sign is pointing. This method works for a wider variety of inequalities, including quadratic inequalities. And in the next section, we'll also use the two-dimensional variation of this method to solve inequalities in two variables.

Here's how you do it:

EXAMPLE:
Find all the solutions to $2 > 8 - \frac{x}{3}$.

SOLUTION

1. Change the inequality into an equation by replacing the inequality sign with an equal sign.

$$2 = 8 - \frac{x}{3}$$

2. Solve the equation.

$$2 = 8 - \frac{x}{3}$$

$$2 - 8 = 8 - \frac{x}{3} - 8$$

$$-6 = -\frac{x}{3}$$

$$-3(-6) = -3\left(-\frac{x}{3}\right)$$

$$18 = x$$

3. Plot the solution to the equation on the number line. If there's more than one solution, plot all of them.

4. The solution to the equation breaks up the number line into two rays—one to the left, the other to the right—with the solution as the boundary. Pick a point in the interior of one of the rays; 0 is a great choice unless it's the boundary point. Plug it into the inequality. If you get a *true* statement, the ray containing that point is the solution ray; if you get a *false* statement, the *other* ray is the solution ray.

   If there's more than one solution, you have two rays and one or more intervals. Test a point in each section to see if that section is included in the solution set.

The point 18 divides the number line into two rays, the points to the right and the points to the left. Choose one point to test—say, 0. Plugging 0 into the inequality gives $2 > 8 - \dfrac{0}{3}$, which is a false statement. So the solutions to the inequality lie to the right of 8, on the other side of 0.

5. If the original inequality sign is ≤ or ≥, the boundary point is included in the solution set; if the original inequality sign is < or >, the boundary point is not included.

   If there's more than one boundary point, it may be necessary to test each one individually.

   Since the original inequality sign is >, the boundary point is *not* included:

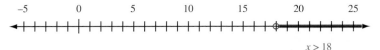

$$x > 18$$

   The solution set is $\{x : x > 18\}$.

Try this next example on your own before reading the solution.

EXAMPLE: Solve with boundary

Solve $6x - 11 \leq 10$

SOLUTION

Change to an equation and solve:

$$6x - 11 = 10$$
$$6x - 11 + 11 = 10 + 11$$
$$6x = 21$$
$$\frac{6x}{6} = \frac{21}{6}$$
$$x = \frac{7}{2}$$

Plot the solution $x = \frac{7}{2}$ on the number line.

Test a point; 0 will do. Plugging 0 into the inequality gives $6(0) - 11 \leq 10$, which is a true statement. So the solution ray is the one that includes 0, to the left of $\frac{7}{2}$. Since the original inequality sign is $\leq$, the boundary point is included.

The solution set is $\{x : x \leq \frac{7}{2}\}$.

Note that the original inequality sign doesn't tell us anything about the direction of the sign in the solution. (We get the direction of the solution sign by testing a point and picking the correct ray, to the left or to the right of the boundary point.) The original inequality sign only tells us whether the solution set includes the boundary point.

Inequalities in Two Variables

Inequalities in two variables are most often solved using a graph in the coordinate plane. Their solutions are most often regions in the plane.

The method of solution is essentially the same as the one introduced in the last section for one-variable inequalities:

1. Convert the inequality into an equation.

2. Graph the equation.

3. The graph—a line or a curve—will divide the plane into two regions. One of these is the solution region; the other is not. Pick a point and test it in the inequality to determine which is which. It usually makes sense to use the origin—unless the graph passes through it, in which case you should use a different point.

4. The graph of the equation is included if the original inequality sign is ≤ or ≥. Use a solid line to indicate this. The graph is not included if the original inequality sign is < or >; use a dashed line to indicate this.

EXAMPLE: Closed half-plane

Identify the region in the plane where $y \leq 3x - 1$.

SOLUTION

Convert the inequality into an equation and graph it: $y = 3x - 1$ is a straight line with slope 3 and y-intercept –1. Since the inequality sign is $\leq$, use a solid line.

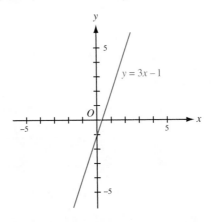

The origin is not on the line, so it will do as a test point: $0 \leq 3(0) - 1$ is a false statement, so $(0, 0)$ is not in the solution region; the solution region is below the line.

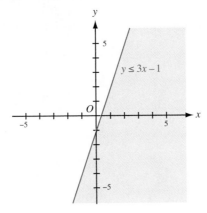

Above or Below the Line? If the inequality is in $y >$ or $y \geq$ form, the solution region is above the line or curve; if the inequality is in $y <$ or $y \leq$ form, the solution region is below.

In the example above, the inequality is in $y \leq$ form, so the solution region is below the line.

This will work even for inequalities that are not linear.

Try this example on your own before reading the solution.

EXAMPLE: Open half-plane
Graph the inequality $y > \dfrac{-x + 1}{2}$.

SOLUTION
Convert the inequality into an equation and graph it:

$y = \dfrac{-x + 1}{2}$ is a straight line with slope $-\dfrac{1}{2}$ and y-intercept $\dfrac{1}{2}$.

Since the inequality sign is >, use a dashed line.

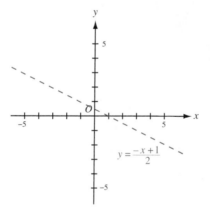

$$y = \frac{-x + 1}{2}$$

The origin is not on the line, so use it as a test point: $0 > \dfrac{-0+1}{2}$ is false, so (0, 0) is not in the solution region; the solution region is above the line.

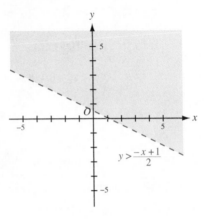

Summary

There are three ways of representing segments of the real number line, which corresponds to three different ways of representing solutions to inequalities:

1. Set notation: $\{x : a \le x < b\}$

2. Interval notation: $[a, b)$

3. Graph:

Set	Interval	Graph	Open? Closed?
$\{x : x < a\}$	$(-\infty, a)$		open
$\{x : x > a\}$	(a, ∞)		open
$\{x : x \le a\}$	$(-\infty, a]$		closed
$\{x : x \ge a\}$	$[a, \infty)$		closed
$\{x : a < x < b\}$	(a, b)		open
$\{x : a \le x \le b\}$	$[a, b]$		closed
$\{x : a \le x < b\}$	$[a, b)$		half-open
$\{x : a < x \le b\}$	$(a, b]$		half-open
$\{x : x \ne a\}$	$(-\infty, a) \cup (a, \infty)$		open
$\{a\}$	$[a, a]$		closed
all real numbers	$(-\infty, \infty)$		open, closed
$\varnothing$			open, closed

Manipulating inequalities algebraically

1. **Adding and subtracting:** You can add or subtract the same number from both sides of an inequality: $a < b$ is equivalent to $a \pm c < b \pm c$ for any c.

2. **Multiplying and dividing by positive number:** You can multiply or divide both sides of an inequality by any positive number: $a < b$ is equivalent to $ac < bc$ for positive c.

3. **Multiplying and dividing by negative number:** You must flip the inequality sign when multiplying or dividing by a negative number: $a < b$ is equivalent to $ac > bc$ for negative c.

Solving inequalities graphically

This general method works for many types of inequalities, including one- and two-variable linear and quadratic inequalities.

1. Change the inequality to an equation.

2. Solve the equation.

3. Graph all solutions to the equation. Plot solutions to one-variable equations on the number line, to two-variable inequalities on the coordinate plane.

4. Test a point in every region and a point on every boundary in the inequality. If a point in a region satisfies the inequality, the whole region does; if it doesn't, then the region doesn't.

Sample Test Questions

Answers to these questions begin on page 410.

1. Write down an inequality in x that corresponds to each graph.

   A.

   B.

   C.

   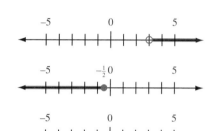

2. Express the set represented by each graph in interval notation.

   A.

   B.

   C.

   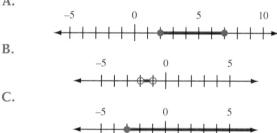

3. Draw the graph of each inequality on a number line.

   A. $x < 2$

   B. $x \geq -4$

   C. $x > 7.5$

   D. $-3 \leq x \leq 3$

   E. $-1 < x \leq 2$

   F. $-3.5 < x < -1.5$

4. Draw the set represented by each interval on the number line.

   A. $[0, 2)$

   B. $(-3, 1.5)$

   C. $[0.25, 1]$

   D. $(-\infty, 5)$

5. Solve each inequality algebraically. Give your answer in set notation or in interval notation.

   A. $x + 3 \leq 7$

   B. $\frac{1}{2}x > 30$

   C. $9 \leq -3x \leq 21$

   D. $-4x + 1 > 7$

   E. $8 > 3x - 1 \geq 2$

   F. $-2x + 7 \leq 4x - 5$

6. Solve and graph each inequality.

   A. $4x + 11 < 15$

   B. $-9 > -2x + 3$

   C. $9x + 7 \leq 12 - x$

   D. $x \geq x - 1$

7. Solve each compound inequality.

   A. $x + 2 > 4 - x > 1$

   B. $3 - x < -2 \leq 1 - x$

   C. $7 - x \leq 5 \leq 9 + x$

   D. $2x - 2 \geq 4 \geq x + 1$

8. Translate each sentence into an inequality using the variable defined in parentheses.

   A. Lily's coffee cup is less than half-full. (Let c be the fraction of coffee in Lily's cup.)

   B. Maximilian has read at least 4 novels by Toni Morrison. (Let b be the number of Toni Morrison novels that Maximilian has read.)

   C. Vladimir owns 12 goldfish, which is no more than the number of goldfish that Boris owns. (Let g be the number of goldfish that Boris owns.)

   D. Keith estimates that he has at least 3 and at most 7 pen pals. (Let p be the number of pen pals that Keith has.)

   E. Brooklyn makes $8 per hour working checkout at the library. She makes more than $70 each week. (Let h be the number of hours that Brooklyn works per week.)

9. David is flying across the country to visit his parents. He has brought his laptop and a spare laptop battery and plans to watch episodes of *Arrested Development*, his favorite TV show, during the flight. Each of his two batteries lasts 1 hour and 31 minutes playing video, and each episode is 21 minutes long.

   A. Let x represent the number of episodes that David can watch during the flight. Write down an inequality in x.

   B. Solve this inequality.

   C. At most, how many episodes of *Arrested Development* will David be able to see during the flight?

10. A triangle has sides of lengths 8 and 13. What are the possible lengths for the third side? *Hint:* Use the **triangle inequality**: the sum of the lengths of any two sides of a triangle always exceeds the third.

11. Graph the region of solutions to each inequality in the xy-plane.

    A. $y < 2x - 1$

    B. $2x + 3y \geq 6$

12. Graph the region of solutions to the system of inequalities.

$$y > \frac{1}{2}x - 2$$

$$y > -3x + 5$$

13. Graph the equation $y = x^2$ in the xy-plane. Use the graph to shade the region of solutions to $y \geq x^2$.

14.

  A. Graph the region of solutions to $y < x^2 - 4$ in the xy-plane.

  B. Look at the x-axis on the graph and use it to write down the solution set to the one-variable inequality $0 < x^2 - 4$.

Rational Expressions

9

Overview

A rational expression is just like a fraction, except that the numerator and the denominator can be polynomials instead of just regular numbers. The good news is that working with rational expressions is pretty much just like working with fractions. You reduce them by canceling common factors, you add them by first converting to common denominators, and you multiply them by multiplying numerators and denominators.

The bad news is that rational expressions, though not difficult conceptually, can be quite cumbersome to deal with. So many terms, so many coefficients, so many pluses and minuses—it's very easy to make a silly arithmetic mistake, even for the most meticulous math students. The best advice is to do plenty of spot checks: pick a number and plug it in for the variable everywhere. If you did everything right, you should get the same values on both sides of your equation.

Introducing Rational Expressions

A **rational expression** is a fraction where the numerator and the denominator are both polynomials. All of these are rational expressions:

$$\frac{3x^4 - x}{2x + 1} \qquad \frac{8ab}{a - b} \qquad \frac{4}{y^3} \qquad \frac{x^2(x + 1)(x - 8)}{-3x^2 - x - 1}$$

TRICKY POINTS

Rational Expressions We say that a rational expression is any *quotient of two polynomials.* This is true, but can be a little misleading. For example, $2x - 1$ is also a rational expression because it can be written as $\dfrac{2x - 1}{1}$.

To be precise, a rational expression is an expression that *can be written* as a quotient of two polynomials. So a number like -1.3 is a rational expression: it can be written as, for example, $\dfrac{-13}{10}$, and both -13 and 10 are polynomials, even if they don't involve any variables.

EXAMPLE: Identifying rational expressions

Which of the following are rational expressions?

a. $\dfrac{4x^3 - 2x^2 + 1}{x^2 - 1}$

b. $\dfrac{1}{2}x^3$

c. $\dfrac{-y}{(x - 1)(x + 3)}$

d. $\dfrac{x + y}{\sqrt{y}}$

e. $\dfrac{2^x}{x^2}$

f. $\dfrac{x^{-3}}{2x + 1}$

SOLUTION

a. Rational expression: the quotient of one polynomial $(4x^3 - 2x^2 + 1)$ by another $(x^2 - 1)$.

b. Rational expression: can be expressed as the quotient of two polynomials (for example, $\dfrac{x^3}{2}$).

c. Rational expression: the quotient of two polynomials (namely, $-y$ and $(x - 1)(x + 3)$).

d. Not a rational expression: the denominator is $\sqrt{y}$, which is not a polynomial.

e. Not a rational expression: the numerator is 2^x, which has a variable exponent and so is not a polynomial.

f. Rational expression: can be rewritten as $\dfrac{x^{-3}}{2x + 1} = \dfrac{1}{x^3(2x + 1)}$, which is the quotient of two polynomials (1 and $x^3(2x + 1)$).

Evaluating Rational Expressions

Like any other expression, a rational expression may take on different values depending on the values that one plugs in for the variables. For example, if $x = 2$, then the rational expression $\frac{x+1}{x-3}$ takes on the value $\frac{(2)+1}{(2)-3} = -3$.

EXAMPLE: Evaluating rational expressions

Evaluate each rational expression for the variable values given.

a. $\frac{4x^2 - 7x + 5}{x + 2}$ for $x = 3$

b. $\frac{m^2 - n^3}{2n}$ for $(m, n) = (-10, 4)$

c. $\frac{x - 1}{x - 2}$ for $x = 2$

SOLUTION

a. Plug in $x = 3$:

$$\frac{4x^2 - 7x + 5}{x + 2} = \frac{4(3)^2 - 7(3) + 5}{(3) + 2} = \frac{20}{5} = 4$$

b. Plug in $m = -10$ and $n = 4$:

$$\frac{m^2 - n^3}{2n} = \frac{(-10)^2 - (4)^3}{2(4)} = \frac{36}{8} = \frac{9}{2}$$

c. Plug in $x = 2$:

$$\frac{x - 1}{x - 2} = \frac{(2) - 1}{(2) - 2} = \frac{1}{0}$$

The expression $\frac{1}{0}$ is not defined: you cannot divide by 0. And there's no way around this problem: the denominator of $\frac{x-1}{x-2}$ is $x - 2$, which equals zero for $x = 2$. The rational expression $\frac{x-1}{x-2}$ simply doesn't make sense when $x = 2$: you cannot evaluate it.

We say that $\frac{x-1}{x-2}$ is **undefined** when $x = 2$.

ZERO DENOMINATOR

Rational expressions only make sense *when the denominator is not equal to zero*. When the denominator is equal to zero, they are undefined.

It is very important to be able to take a rational expression and determine for which values of the variable(s) the expression is defined. Fortunately, a rational expression is defined almost all the time.

For example, take a look at:

$$\frac{x^2 - 1}{x + 3}$$

You can plug in $x = 0$, $x = 1$, $x = -\frac{3}{2}$, and so on, and evaluate it to get a number. There's only one problem point, $x = -3$, because that's exactly the value that makes the denominator $x + 3$ equal to zero. We can say that $\frac{x^2 - 1}{x + 3}$ is defined for all real numbers except –3, or whenever $x \neq -3$.

KEY POINTS

Defined Rational Expression A rational expression is defined whenever its denominator is nonzero. In other words, if P and Q are polynomials, then the rational expression

$$\frac{P}{Q}$$

is defined whenever Q does not evaluate to 0.

EXAMPLE: Determining when a rational expression is defined

When is each of these rational expressions defined?

a. $\dfrac{z+3}{2z-5}$

b. $\dfrac{-x+7}{(x-1)(3x+1)}$

c. $\dfrac{3y^2-4}{9}$

d. $\dfrac{3}{x^2-6x-16}$

e. $\dfrac{x-5}{x-5}$

f. $\dfrac{7b-5}{b^2+1}$

g. $\dfrac{1}{x-y}$

SOLUTION

a. The expression is defined unless the denominator equals zero: that is, unless $2z - 5 = 0$. Solve:

$$2z - 5 = 0$$
$$2z - 5 + 5 = 0 + 5$$
$$2z = 5$$
$$z = \frac{5}{2}$$

So $\dfrac{z+3}{2z-5}$ is defined for all values of z except $z = \dfrac{5}{2}$.

b. The expression is defined unless the denominator equals zero: that is, unless $(x - 1)(3x + 1) = 0$. To solve, use the zero product property: if $(x - 1)(3x + 1) = 0$, then $x - 1 = 0$ or $3x + 1 = 0$:

$$x - 1 = 0 \quad \text{or} \quad 3x + 1 = 0$$
$$\downarrow \qquad\qquad \downarrow$$
$$x = 1 \quad \text{or} \quad x = -\frac{1}{3}$$

So $\dfrac{-x+7}{(x-1)(3x+1)}$ is defined for all values of x except $x = 1$ and $x = -\dfrac{1}{3}$.

c. The expression is defined unless the denominator equals zero. But the denominator is 9, which never equals 0 regardless of the value of y. So $\dfrac{3y^2 - 4}{9}$ is defined for all real values of y.

d. The expression is defined unless the denominator equals zero: that is, unless $x^2 - 6x - 16 = 0$. To find out when that happens, factor $x^2 - 6x - 16$. Find two numbers whose product is -16 and whose sum is -6. The factor pairs of -16 are -1 and 16, -2 and 8, -4 and 4, -8 and 2, and -16 and 1; by trial and error we see that -8 and 2 work:

$$x^2 - 6x - 16 = (x - 8)(x + 2)$$

So the expression is defined unless $x^2 - 6x - 16 = (x - 8)(x + 2) = 0$. By the zero product property, $(x - 8)(x + 2) = 0$ means that $x - 8 = 0$ or $x + 2 = 0$:

$$x - 8 = 0 \quad \text{or} \quad x + 2 = 0$$
$$\downarrow \qquad\qquad\qquad \downarrow$$
$$x = 8 \quad \text{or} \quad x = -2$$

So $\dfrac{3}{x^2 - 6x - 16} = 0$ is defined for all values of x except $x = 8$ and $x = -2$.

e. The expression is defined unless the denominator equals zero: that is, unless $x - 5 = 0$, which happens when $x = 5$. So $\dfrac{x - 5}{x - 5}$ is defined for all values of x except $x = 5$.

f. The expression is defined unless the denominator equals zero: that is, unless $b^2 + 1 = 0$. But this never happens: b^2 is greater than or equal to 0 for all values of b, so $b^2 + 1$ is greater than or equal to 1 for all values of b.

So $\dfrac{7b - 5}{b^2 + 1}$ is defined for all values of b.

If you're feeling enterprising, you can verify that $b^2 + 1$ has no roots by evaluating its discriminant, which is the square of the b coefficient minus 4 times the product of the b^2 coefficient and the constant term. The b coefficient is 0, the b^2 and the constant coefficients are both 1, so

$$\text{discriminant is } (0)^2 - 4(1)(1) = -4$$

Since the discriminant is negative, the polynomial $b^2 + 1$ has no real roots; it doesn't equal zero for any real value of b.

g. The expression is defined unless the denominator equals zero: that is, unless $x - y = 0$. This happens only when $x = y$, and that's the best that we can say. So $\dfrac{1}{x - y}$ is defined for all values of x and y except if $x = y$.

Simplifying Rational Expressions

EQUIVALENT FRACTIONS AND CANCELLATION

Two fractions are **equivalent** if they represent the same quantity. For example, $\dfrac{2}{3}$ and $\dfrac{6}{9}$ are equivalent fractions: both identify the same point on the number line.

To convert a fraction to an equivalent fraction with a different denominator, multiply or divide both the numerator and the denominator of the fraction by the same number. For example,

$$\frac{2}{5} = \frac{2 \cdot 3}{5 \cdot 3} = \frac{6}{15}$$

and

$$\frac{40}{60} = \frac{40 \div 10}{60 \div 10} = \frac{4}{6}$$

When dividing, you may choose to use **cancellation** notation rather than writing everything out. If you spot that the numerator and the denominator have a common factor, cross them out, divide in your head, and note what's left over:

$$\frac{40}{60} = \frac{\overset{4}{\cancel{40}}}{\underset{6}{\cancel{60}}} = \frac{4}{6}$$

EQUIVALENT RATIONAL EXPRESSIONS AND CANCELLATION

Two rational expressions are **equivalent** if, when you plug in the same value of x, both expressions give you the same value. This has to work for all values of x: if one expression is undefined for a particular value of x, the other one must be undefined also.

For example, $\frac{2(x-5)}{3(x-5)}$ and $\frac{2}{3}$ are *almost* equivalent rational expressions. Whenever you evaluate $\frac{2(x-5)}{3(x-5)}$ for some value of x, you get a fraction whose numerator is two-thirds its denominator, which means that this fraction evaluates to $\frac{2}{3}$. And that's exactly what you get when you evaluate the rational expression $\frac{2}{3}$ for that value of x. (Since the rational expression $\frac{2}{3}$ doesn't depend on x, it evaluates to $\frac{2}{3}$ no matter what value of x you use.) Take a look: plugging $x = 10$ into $\frac{2(x-5)}{3(x-5)}$ gives $\frac{2((10)-5)}{3((10)-5)} = \frac{10}{15} = \frac{2}{3}$.

However, this doesn't work for $x = 5$: When you evaluate $\frac{2(x-5)}{3(x-5)}$ for $x = 5$, you get something undefined. But when you evaluate the rational expression $\frac{2}{3}$ for $x = 5$, you still get the $\frac{2}{3}$. So to make a rational expression that is equivalent to $\frac{2(x-5)}{3(x-5)}$, we qualify $\frac{2}{3}$ with a special note:

$$\frac{2(x-5)}{3(x-5)} \text{ is equivalent to } \frac{2}{3}, \quad x \neq 5$$

To convert a rational expression to an equivalent rational expression, multiply or divide both the numerator and the denominator by the same polynomial:

$$\frac{2x+1}{x-3} = \frac{x(2x+1)}{x(x-3)} = \frac{2x^2+x}{x^2-3x}, \quad x \neq 0, 3$$

or

$$\frac{(x+2)(x-2)}{(x+3)(x-2)} = \frac{(x+2)(x-2) \div (x-2)}{(x+3)(x-2) \div (x-2)} = \frac{x+2}{x+3}, \quad x \neq -3, 2$$

You can also use cancellation notation with rational expressions:

$$\frac{(x+2)(x-2)}{(x+3)(x-2)} = \frac{(x+2)\cancel{(x-2)}}{(x+3)\cancel{(x-2)}} = \frac{x+2}{x+3}, \quad x \neq -3, 2$$

FRACTIONS

A fraction is said to be in **simplified**, or **reduced**, form, also known as **lowest terms**, if its numerator and denominator have no common factors (1 and –1 don't count). For example, $\frac{1}{2}$ is in lowest terms, but $\frac{9}{18}$ is not, as both 9 and 18 are divisible by 3 and by 9.

There are at least a couple of good methods for reducing a fraction to lowest terms; we'll just review the one that's most useful for rational expressions. Factor both the numerator and the denominator completely, into prime factors. Then eliminate common factors.

EXAMPLE: Lowest terms

Express $\frac{12}{18}$ in lowest terms.

SOLUTION

Factor the numerator and the denominator:

$$12 = 2 \cdot 2 \cdot 3 \qquad 18 = 2 \cdot 3 \cdot 3$$

Reduce by eliminating common factors:

$$\frac{12}{18} = \frac{2 \cdot 2 \cdot 3}{2 \cdot 3 \cdot 3} = \frac{\cancel{2} \cdot 2 \cdot 3}{\cancel{2} \cdot 3 \cdot 3} = \frac{2 \cdot 3}{3 \cdot 3} = \frac{2 \cdot \cancel{3}}{\cancel{3} \cdot 3} = \frac{2}{3}$$

Since 2 and 3 have no common factors, $\frac{12}{18}$ reduced to lowest terms becomes $\frac{2}{3}$.

RATIONAL EXPRESSIONS

A rational expression is said to be in *simplified* or *reduced form* if its numerator and denominator have no common polynomial factors. Again, factors of 1 and –1 don't count.

For example, $\dfrac{x(x+1)}{3x-2}$ is in simplified form. But $\dfrac{(x+1)^2}{(3x-2)(x+1)}$ is not in simplified form: both the numerator and the denominator have a factor of $(x+1)$.

To simplify a rational expression, first factor both the numerator and the denominator—completely. Then eliminate common factors. The result should be a rational expression equivalent to the original but in simplified form, without common factors in the numerator and denominator.

EXAMPLE: Simplifying I

Simplify $\dfrac{(x+1)^2}{(3x-2)(x+1)}$.

SOLUTION

Both the numerator and the denominator are already in factored form, so there's no need to factor further. The only thing to do is eliminate the common factor of $(x+1)$ from both top and bottom.

$$\frac{(x+1)^2}{(3x-2)(x+1)} = \frac{(x+1)(x+1)}{(3x-2)(x+1)}$$

$$= \frac{\cancel{(x+1)}(x+1)}{(3x-2)\cancel{(x+1)}}$$

$$= \frac{x+1}{(3x-2)}, \qquad x \neq -1, \frac{2}{3}$$

Try these on your own before reading through the solutions.

EXAMPLE: Simplifying II

Write each rational expression in simplified form.

a. $\dfrac{x(x-3)}{x^2(x-3)^3}$

b. $\dfrac{12x-48}{40(x-5)}$

c. $\dfrac{x^2-25}{x^2-10x+25}$

d. $\dfrac{x+2}{x^2+2x+4}$

e. $\dfrac{2x^2-2x-24}{2x^2+x-15}$

f. $\dfrac{x^4-81}{4x^2+9x-9}$

SOLUTION

a. $\dfrac{x(x-3)}{x^2(x-3)^3} = \dfrac{1}{x(x-3)^2}, \quad x \neq 3, 0$

Both the numerator and the denominator are already completely factored; all that remains is to eliminate the common factors of x and $x - 3$ from both top and bottom:

$$\frac{x(x-3)}{x^2(x-3)^3} = \frac{x(x-3)}{x \cdot x(x-3)(x-3)^2}$$

$$= \frac{\cancel{x}\cancel{(x-3)}}{\cancel{x} \cdot x\cancel{(x-3)}(x-3)^2}$$

$$= \frac{1}{x(x-3)^2}, \quad x \neq 3, 0$$

b. $\dfrac{12x-48}{40(x-5)} = \dfrac{3(x-4)}{10(x-5)}, \quad x \neq 5$

The denominator is completely factored, but the numerator is not: both $12x$ and -48 are divisible by 12. Pull out common factors to express the numerator as $12x - 48 = 12(x) - 12(4) = 12(x-4)$.

So the rational expression may be rewritten as $\dfrac{12(x-4)}{40(x-5)}$ in factored form.

The only factors common to both the numerator and the denominator are in the number coefficients. Both the 12 in the numerator and the 40 in the denominator are divisible by 4:

$$\frac{12(x-4)}{40(x-5)} = \frac{\overset{3}{\cancel{12}}(x-4)}{\underset{10}{\cancel{40}}(x-5)}$$

$$= \frac{3(x-4)}{10(x-5)}, \qquad x \neq 5$$

c. $\dfrac{x^2-25}{x^2-10x+25} = \dfrac{x+5}{x-5}, \quad x \neq 5$

First, factor the numerator: $x^2 - 25$ is a difference of squares that factors as $(x+5)(x-5)$.

Next, factor the denominator: $x^2 - 10x + 25$ happens to be a perfect square in $a^2 - 2ab + b^2$ form. If you don't notice that it's a perfect square, factor as usual: find two signed numbers whose product is 25 and whose sum is –10. The only pair of numbers that works is –5 and –5, so $x^2 - 10x + 25 = (x - 5)^2$.

Finally, eliminate common factors:

$$\frac{x^2-25}{x^2-10x+25} = \frac{(x+5)(x-5)}{(x-5)(x-5)}$$

$$= \frac{(x+5)\cancel{(x-5)}}{\cancel{(x-5)}(x-5)}$$

$$= \frac{x+5}{x-5}, \qquad x \neq 5$$

d. $\dfrac{x+2}{x^2+2x-4}$ is in simplified form

The numerator is already in factored form, so all that's left is to try to factor the denominator. We look for two signed numbers whose product is –4 and whose sum is 2. The pairs of numbers whose product is –4 are 1 and –4, 2 and –2, and 4 and –1:

pair with product –4	signed sum
1 and –4	–3
2 and –2	0
4 and –1	3

None of these sums is 2, so there is no pair of integers whose product is –4 and whose sum is 2. This means that $x^2 + 2x - 4$ cannot be factored in any helpful way. So the numerator and denominator in $\dfrac{x + 2}{x^2 + 2x - 4}$ are completely factored. Since the numerator and the denominator do not have factors in common, $\dfrac{x + 2}{x^2 + 2x - 4}$ is already in simplified form.

e. $\dfrac{2x^2 - 2x - 24}{2x^2 + x - 15} = \dfrac{2(x - 4)}{2x - 5}, \quad x \neq -3, \dfrac{5}{2}$

First, factor the numerator. Pull out the common factor of 2 to express $2x^2 - 2x - 24 = 2(x^2 - x - 12)$. Now factor $x^2 - x - 12$: find two numbers whose product is –12 and whose sum is –1. By trial and error, 3 and –4 work. So $2(x^2 - x - 12) = 2(x + 3)(x - 4)$.

Next, factor the denominator. The product of the first and last coefficient is $2 \cdot (-15) = -30$, so we look for two numbers whose product is –30 and whose sum is 1. By trial and error, 6 and –5 work. Now rewrite x as $6x - 5x$ and factor by grouping:

$$\begin{aligned}
2x^2 + x - 15 &= 2x^2 + 6x - 5x - 15 \\
&= 2x(x + 3) - 5(x + 3) \\
&= (2x - 5)(x + 3)
\end{aligned}$$

Finally, look at the factored rational expression and eliminate common factors:

$$\begin{aligned}
\dfrac{2x^2 - 2x - 24}{2x^2 + x - 15} &= \dfrac{2(x + 3)(x - 4)}{(2x - 5)(x + 3)} \\
&= \dfrac{2\cancel{(x + 3)}(x - 4)}{(2x - 5)\cancel{(x + 3)}} \\
&= \dfrac{2(x - 4)}{2x - 5}, \qquad x \neq -3, \dfrac{5}{2}
\end{aligned}$$

Note that you have to exclude $x = -3$ even though the simplified form doesn't have a factor of $x + 3$ in the denominator. That's because the original rational expression is not defined for $x = -3$, and you want the simplified form to be equivalent to the original form.

f. $\dfrac{x^4 - 81}{4x^2 + 9x - 9} = \dfrac{(x^2 + 9)(x - 3)}{4x - 3}, \quad x \neq -3, \dfrac{3}{4}$

First, factor the numerator: $x^4 - 81$ is a difference of squares:

$$x^4 - 81 = (x^2)^2 - 9^2$$
$$= (x^2 + 9)(x^2 - 9)$$

At this point, $x^2 + 9$ is a simple sum of squares, so it cannot be factored further. But (tricky!) $x^2 - 9$ is again a difference of squares, so it can be factored once more:

$$(x^2 + 9)(x^2 - 9) = (x^2 + 9)(x + 3)(x - 3)$$

Next, factor the denominator. The product of the first and last coefficients of $4x^2 + 9x - 9$ is $4 \cdot (-9) = -36$, so we need to find a pair of numbers whose product is -36 and whose sum is 9. By trial and error, 12 and -3 work. So rewrite $9x = 12x - 3x$ and factor by grouping:

$$4x^2 + 9x - 9 = 4x^2 + 12x - 3x - 9$$
$$= 4x(x + 3) - 3(x + 3)$$
$$= (4x - 3)(x + 3)$$

Finally, rewrite the rational expression in factored form and cancel factors:

$$\frac{x^4 - 81}{4x^2 + 9x - 9} = \frac{(x^2 + 9)(x + 3)(x - 3)}{(4x - 3)(x + 3)}$$
$$= \frac{(x^2 + 9)\cancel{(x + 3)}(x - 3)}{(4x - 3)\cancel{(x + 3)}}$$
$$= \frac{(x^2 + 9)(x - 3)}{4x - 3}, \quad x \neq -3, \frac{3}{4}$$

RECOGNIZING FACTORS OF –1

Rational expressions can be trickier than fractions to simplify to lowest terms. It's always easy to see when two integers differ by a factor of –1, like 49 and –49: you just look at them. But it can be difficult to recognize that two polynomials differ by a factor of –1. For example, $4x + 7$ and $-4x - 7$ differ by a factor of –1 because

$$-1(4x + 7) = -1(4x) + (-1)(7) = -4x - 7$$

Also, $2x - 3$ and $3 - 2x$ differ by a factor of –1 because

$$-1(2x - 3) = -1(2x) - (-1)(3) = -2x + 3 = 3 - 2x$$

So, a rational expression like $\dfrac{2x - 3}{3 - 2x}$ is not in lowest terms, though both the numerator and the denominator are completely factored (factors of –1 don't count). You can see how to reduce it if you first factor out a –1 from either the top or the bottom:

$$\frac{2x - 3}{3 - 2x} = \frac{2x - 3}{-1(2x - 3)} = \frac{\cancel{2x - 3}}{-1(\cancel{2x - 3})} = \frac{1}{-1} = -1 \text{ , } x \neq \frac{3}{2}$$

TRICKY POINTS

Overrecognizing Beware of overrecognizing. For example, $x + 5$ and $x - 5$ do not differ by a factor of –1; they're two different polynomials. So a rational expression like $\dfrac{x + 5}{x - 5}$ is in simplified form.

Try these on your own before reading through the solutions.

EXAMPLE: Simplifying III: Tricky factors of –1

Simplify each rational expression.

a. $\dfrac{-x-4}{x-4}$

b. $\dfrac{-x+4}{x-4}$

c. $\dfrac{x^2-16}{8-2x}$

SOLUTION

a. $\dfrac{-x-4}{x-4}$ is in simplified form

The numerator and the denominator do not have any factors in common. (The numerator can be reexpressed as $-1(x+4)$ if you wish, but that doesn't help you simplify.)

b. $\dfrac{-x+4}{x-4} = -1, \quad x \neq 4$

The numerator and the denominator differ by a factor of –1:

$$\frac{-x+4}{x-4} = \frac{-1(x-4)}{x-4}$$
$$= \frac{-1\cancel{(x-4)}}{\cancel{x-4}}$$
$$= -1, \quad x \neq 4$$

c. $\dfrac{x^2-16}{8-2x} = \dfrac{x+4}{-2}, \quad x \neq 4$

First, factor the numerator: $x^2 - 16 = (x+4)(x-4)$.

Next, factor the denominator by pulling out a 2: $8 - 2x = 2(4 - x)$.

Rewrite the rational expression; it is now completely factored:

$$\frac{x^2-16}{8-2x} = \frac{(x+4)(x-4)}{2(4-x)}$$

To simplify, note that the $x - 4$ in the numerator and the $4 - x$ in the denominator differ by a factor of -1. Pull it out and cancel:

$$\frac{(x+4)(x-4)}{2(4-x)} = \frac{(x+4)(x-4)}{2(-1)(x-4)}$$

$$= \frac{(x+4)\cancel{(x-4)}}{2(-1)\cancel{(x-4)}}$$

$$= \frac{x+4}{-2}, \qquad x \neq 4$$

Multiplying Rational Expressions

FRACTIONS

To multiply two fractions, we multiply their numerators and multiply their denominators.

> **Multiplying Fractions**
> If a, b, c, and d are numbers with $b, d \neq 0$, then
> $$\frac{a}{b} \cdot \frac{c}{d} = \frac{ac}{bd}$$

EXAMPLE: Multiplying Fractions

Find $\dfrac{4}{9} \cdot \dfrac{15}{22}$.

SOLUTION

Multiply the numerators and multiply the denominators:

$$\frac{4}{9} \cdot \frac{15}{22} = \frac{4 \cdot 15}{9 \cdot 22}$$

We could multiply everything out right now, but if we want to simplify we'd just have to factor everything again. Instead we reduce at the same time:

$$\frac{4 \cdot 15}{9 \cdot 22} = \frac{\overset{2}{\cancel{4}} \cdot 15}{9 \cdot \underset{11}{\cancel{22}}}$$

$$= \frac{2 \cdot \overset{5}{\cancel{15}}}{\underset{3}{\cancel{9}} \cdot 11}$$

$$= \frac{2 \cdot 5}{3 \cdot 11} = \frac{10}{33}$$

RATIONAL EXPRESSIONS

To multiply two rational expressions, multiply their numerators and multiply their denominators.

> **Multiplying Rational Expressions**
>
> If A, B, C, and D are polynomials and neither B nor D is the zero polynomial, then
>
> $$\frac{A}{B} \cdot \frac{C}{D} = \frac{AC}{BD}$$

When multiplying rational expressions, it's best to keep the numerators and the denominators in factored form until all the common factors of the product have been eliminated.

EXAMPLE: Multiplying rational expressions

a. Find the product of $\dfrac{2(x+8)}{x^2+4}$ and $\dfrac{(x+8)}{6x}$.

b. Multiply: $\dfrac{x^2-5x-14}{x^2+5x+4} \cdot \dfrac{-x^2-3x+4}{x^2-8x+7}$.

SOLUTION

a. $\dfrac{2(x+8)}{x^2+4} \cdot \dfrac{(x+8)}{6x} = \dfrac{(x+8)^2}{3x(x^2+4)}$

Multiply and simplify:

$$\frac{2(x+8)}{x^2+4} \cdot \frac{(x+8)}{6x} = \frac{\cancel{2}(x+8)(x+8)}{(x^2+4)(\underset{3}{\cancel{6}}x)}$$

$$= \frac{(x+8)^2}{3x(x^2+4)}$$

The x^2+4 in the denominator does not factor, so we're already in lowest terms.

b. $\dfrac{x^2-5x-14}{x^2+5x+4} \cdot \dfrac{-x^2-3x+4}{x^2-8x+7} = \dfrac{-(x+2)}{x+1}$

Before multiplying, factor each numerator and denominator:

$$x^2-5x-14 = (x+2)(x-7)$$
$$x^2+5x+4 = (x+1)(x+4)$$
$$-x^2-3x+4 = -(x+4)(x-1)$$
$$x^2-8x+7 = (x-1)(x-7)$$

Now, multiply and simplify:

$$\frac{x^2-5x-14}{x^2+5x+4} \cdot \frac{-x^2-3x+4}{x^2-8x+7} = \frac{(x+2)(x-7)}{(x+1)(x+4)} \cdot \frac{-(x+4)(x-1)}{(x-1)(x-7)}$$

$$= \frac{(x+2)\cancel{(x-7)}(-1)\cancel{(x+4)}\cancel{(x-1)}}{(x+1)\cancel{(x+4)}\cancel{(x-1)}\cancel{(x-7)}}$$

$$= \frac{-(x+2)}{x+1}$$

Dividing Rational Expressions

FRACTIONS

To divide by a fraction, multiply by the reciprocal of that fraction.

> **Dividing Fractions**
>
> If a, b, c, and d are real numbers with b, c, and d nonzero, then
>
> $$\frac{a}{b} \div \frac{c}{d} = \frac{a}{b} \cdot \frac{d}{c} = \frac{ad}{bc}$$

It's best to convert divisions to multiplications before doing any canceling—otherwise you run the risk of losing track of what belongs in the numerator and what belongs in the denominator.

EXAMPLE: Dividing Fractions

Compute $\dfrac{5}{18} \div \dfrac{2}{15}$.

SOLUTION

Flip the second fraction and multiply:

$$\frac{5}{18} \div \frac{2}{15} = \frac{5}{18} \cdot \frac{15}{2}$$
$$= \frac{5 \cdot \overset{5}{\cancel{15}}}{\underset{6}{\cancel{18}} \cdot 2}$$
$$= \frac{25}{12}$$

RATIONAL EXPRESSIONS

To divide one rational expression by another, multiply the first by the reciprocal of the second.

> **Dividing Rational Expressions**
>
> If A, B, C, and D are polynomials and none of B, C, or D is the zero polynomial, then
>
> $$\frac{A}{B} \div \frac{C}{D} = \frac{A}{B} \cdot \frac{D}{C} = \frac{AD}{BC}$$

EXAMPLE: Dividing rational expressions

a. Divide $\dfrac{5a^2}{6b^5}$ by $\dfrac{30a^4}{b^2}$.

b. Compute $\dfrac{y^2 - 1}{y^2 + y + 1} \div \dfrac{y - 1}{y + 1}$.

SOLUTION

a. $\dfrac{5a^2}{6b^5} \div \dfrac{30a^4}{b^2} = \dfrac{1}{36a^2b^3}$

   Flip the second fraction and multiply:

   $$\frac{5a^2}{6b^5} \div \frac{30a^4}{b^2} = \frac{5a^2}{6b^5} \cdot \frac{b^2}{30a^4}$$

   $$= \frac{\cancel{5} \cdot \cancel{a} \cdot \cancel{a} \cdot \cancel{b} \cdot \cancel{b}}{6\cancel{b} \cdot \cancel{b} \cdot b \cdot b \cdot b \cdot \underset{6}{\cancel{30}} \cdot \cancel{a} \cdot \cancel{a} \cdot a \cdot a}$$

   $$= \frac{1}{36a^2b^3}$$

b. $\dfrac{y^2 - 1}{y^2 + y + 1} \div \dfrac{y - 1}{y + 1} = \dfrac{(y + 1)^2}{y^2 + y + 1}$

   Flip the second fraction and multiply:

   $$\frac{y^2 - 1}{y^2 + y + 1} \div \frac{y - 1}{y + 1} = \frac{(y+1)(y-1)}{y^2 + y + 1} \cdot \frac{y+1}{y-1}$$

   $$= \frac{(y+1)\,\cancel{(y-1)} \cdot (y+1)}{(y^2 + y + 1)\,\cancel{(y-1)}}$$

   $$= \frac{(y+1)^2}{y^2 + y + 1}$$

Adding and Subtracting Rational Expressions

FRACTIONS

To add or subtract two fractions with the same denominator, simply add or subtract their numerators.

> ### *Adding and Subtracting Fractions (Same Denominator)*
>
> If a, b, and c are real numbers with $b \neq 0$, then
> $$\frac{a}{b} + \frac{c}{b} = \frac{a+c}{b}$$
> and
> $$\frac{a}{b} - \frac{c}{b} = \frac{a-c}{b}$$

For example:

$$\frac{7}{9} - \frac{1}{9} = \frac{7-1}{9} = \frac{\overset{2}{\cancel{6}}}{\underset{3}{\cancel{9}}} = \frac{2}{3}$$

To add or subtract two fractions with different denominators, you first have to convert both fractions to equivalent fractions with a *common denominator*. If you're working with $\frac{a}{b}$ and $\frac{c}{d}$, then bd is one reasonable common denominator. The fractions convert to $\frac{a}{b} = \frac{ad}{bd}$ and $\frac{c}{d} = \frac{bc}{bd}$, so now you can add or subtract.

Adding and Subtracting Fractions (Different Denominators)

If a, b, c, and d are real numbers with $b, d \neq 0$, then

$$\frac{a}{b} + \frac{c}{d} = \frac{ad}{bd} + \frac{bc}{bd} = \frac{ad + bc}{bd}$$

and

$$\frac{a}{b} - \frac{c}{d} = \frac{ad}{bd} - \frac{bc}{bd} = \frac{ad - bc}{bd}$$

The product of the two denominators is not always the most efficient common denominator to use. But it is simple and reliable.

EXAMPLE: Adding integer fractions

Compute $\frac{3}{8} + \frac{5}{6}$.

SOLUTION

The product of the two denominators is 48. Convert the two fractions:

$$\frac{3}{8} = \frac{3 \cdot 6}{8 \cdot 6} = \frac{18}{48}$$

and

$$\frac{5}{6} = \frac{5 \cdot 8}{6 \cdot 8} = \frac{40}{48}$$

Add and simplify:

$$\frac{18}{48} + \frac{40}{48} = \frac{58}{48} = \frac{29}{24}$$

RATIONAL EXPRESSIONS

Addition and subtraction work the same way with rational expressions as with integer fractions. If the two rational expressions have the same denominator, then all you have to do is add or subtract their denominators.

Adding and Subtracting Rational Expressions (Same Denominator)

If *A*, *B*, and *C* are polynomials with *B* not the zero polynomial, then

$$\frac{A}{B} + \frac{C}{B} = \frac{A + C}{B}$$

and

$$\frac{A}{B} - \frac{C}{B} = \frac{A - (C)}{B}$$

EXAMPLE: Adding, same denominator

Find $\dfrac{2x - 1}{3x - 7} + \dfrac{3x - 8}{3x - 7}$.

SOLUTION

$$\frac{2x - 1}{3x - 7} + \frac{3x - 8}{3x - 7} = \frac{(2x - 1) + (3x - 8)}{3x - 7}$$

$$= \frac{5x - 9}{3x - 7}$$

When subtracting rational expressions, be very careful with the minus sign. The most popular option is to bring it up into the numerator and distribute very carefully, to each term. Watch how the minus sign is treated in the example below and in all the subtraction problems in this section.

EXAMPLE: Subtracting, same denominator

Find $\dfrac{x^2 + 2x - 7}{x^2 - 25} - \dfrac{x^2 - x + 8}{x^2 - 25}$.

SOLUTION

Remember to put parentheses around the numerator of the second fraction!

$$\frac{x^2 + 2x - 7}{x^2 - 25} - \frac{x^2 - x + 8}{x^2 - 25} = \frac{(x^2 + 2x - 7) - (x^2 - x + 8)}{x^2 - 25}$$

$$= \frac{x^2 + 2x - 7 - x^2 + x - 8}{x^2 - 25}$$

$$= \frac{(x^2 - x^2) + (2x + x) + (-7 - 8)}{x^2 - 25}$$

$$= \frac{3x - 15}{x^2 - 25}$$

The subtraction is done; now check whether the result is in lowest terms. Factor top and bottom and compare factors:

$$\frac{3x - 15}{x^2 - 25} = \frac{3(x - 5)}{(x + 5)(x - 5)}$$

$$= \frac{3\cancel{(x - 5)}}{(x + 5)\cancel{(x - 5)}}$$

$$= \frac{3}{x + 5}, \qquad x \neq \pm 5$$

Both of the examples we've looked at involved two rational expressions with the same denominator. If the denominators are different, you have to convert to equivalent rational expressions with a common denominator before performing the addition or subtraction. As with integer fractions, the product of the two denominators works as a common denominator:

> *Adding and Subtracting Rational Expressions (Different Denominators)*
>
> If A, B, and C are polynomials and neither B nor D is the zero polynomial, then
>
> $$\frac{A}{B} + \frac{C}{D} = \frac{AD}{BD} + \frac{BC}{BD} = \frac{AD + BC}{BD}$$
>
> and
>
> $$\frac{A}{B} - \frac{C}{D} = \frac{AD}{BD} - \frac{BC}{BD} = \frac{AD - (BC)}{BD}$$

EXAMPLE: Adding and subtracting, product denominator

a. Find $\dfrac{3}{x+2y} + \dfrac{5}{x-3y}$.

b. Find $\dfrac{z+2}{z-2} - \dfrac{z-2}{z+2}$.

SOLUTION

a. $\dfrac{3}{x+2y} + \dfrac{5}{x-3y} = \dfrac{8x+y}{(x+2y)(x-3y)}$

Using the common denominator $(x+2y)(x-3y)$, convert the fractions and add:

$$\dfrac{3}{x+2y} + \dfrac{5}{x-3y} = \dfrac{3(x-3y)}{(x+2y)(x-3y)} + \dfrac{5(x+2y)}{(x-3y)(x+2y)}$$

$$= \dfrac{3(x-3y)+5(x+2y)}{(x+2y)(x-3y)}$$

$$= \dfrac{3x-9y+5x+10y}{(x+2y)(x-3y)}$$

$$= \dfrac{8x+y}{(x+2y)(x-3y)}$$

b. $\dfrac{z+2}{z-2} - \dfrac{z-2}{z+2} = \dfrac{8z}{(z+2)(z-2)}$

Using the common denominator $(z-2)(z+2)$, convert the fractions and subtract:

$$\dfrac{z+2}{z-2} - \dfrac{z-2}{z+2} = \dfrac{(z+2)(z+2)}{(z-2)(z+2)} - \dfrac{(z-2)(z-2)}{(z+2)(z-2)}$$

$$= \dfrac{(z+2)^2-(z-2)^2}{(z+2)(z-2)}$$

$$= \dfrac{z^2+4z+4-(z^2-4z+4)}{(z+2)(z-2)}$$

$$= \dfrac{z^2+4z+4-z^2+4z-4}{(z+2)(z-2)}$$

$$= \dfrac{8z}{(z+2)(z-2)}$$

LEAST COMMON MULTIPLE FOR INTEGERS

A **common multiple** of two integers is any number that is divisible by both of them. For example, 80 is a common multiple of 4 and 10.

The **least common multiple (LCM)** of two numbers is just that—the least of their common multiples. The LCM of 4 and 10 is 20.

To find the LCM of two or more integers, first find their prime factorization. Each prime that appears in one of the integers must appear in the LCM. If a prime appears more than once in any integer, it must appear in the LCM the greatest number of times that it appears anywhere. The easiest way to understand the LCM is to look at some examples.

EXAMPLE: LCM of integers

  a. Find the LCM of 42 and 30.

  b. Find the LCM of 24 and 180.

  c. Find the LCM of 10, 15, and 18.

SOLUTION

  a. The LCM of 42 and 30 is 210.
     Factor: $42 = 2 \cdot 3 \cdot 7$ and $30 = 2 \cdot 3 \cdot 5$. Their LCM must contain a 2, a 3, a 5, and a 7. So the LCM is
     $2 \cdot 3 \cdot 5 \cdot 7 = 210$.

  b. The LCM of 24 and 180 is 360.
     Factor: $24 = 2^3 \cdot 3$ and $180 = 2^2 \cdot 3^2 \cdot 5$. The LCM must include factors of 2, 3, and 5. The 2 appears three times in 24 and the 3 appears twice in 180, so the LCM is
     $2^3 \cdot 3^2 \cdot = 360$.

  c. The LCM of 10, 15, and 18 is 90.
     Factor: $10 = 2 \cdot 5$, $15 = 3 \cdot 5$, and $18 = 2 \cdot 3^2$. The LCM must include factors of 2, 3, and 5; since 3 appears twice in 18, it must appear twice in the LCM. So the LCM is
     $2 \cdot 3^2 \cdot 5 = 90$.

LEAST COMMON MULTIPLE FOR POLYNOMIALS

Polynomials, like integers, have common multiples and least common multiples (LCMs). For example, the product of $x^2(x + 3)$ and $x(x + 3)^3(x - 5)$ is $x^3(x + 3)^4(x - 5)$; it is a common multiple of the two polynomials. Their least common multiple is $x^2(x + 3)^3(x - 5)$; it has fewer factors.

The LCM of two or more polynomials is found the same way you find the LCM of integers. Factor each polynomial. The LCM will include each factor that appears in any polynomial, as many times as the maximum number that it appears in any polynomial. Again, it's easiest to see what's going on by looking at some examples.

EXAMPLE: LCM of polynomials

a. Find the LCM of $8x^7y$ and $10xy^3$.

b. Find the LCM of $(3m - 2n)(m + 5)$ and $(2n - 3m)(m + n)$.

c. Find the LCM of $14(a^2 - b^2)$, $21(a^2 + 2ab + b^2)$, and $12(a + b)$.

SOLUTION

a. The LCM of $8x^7y$ and $10xy^3$ is $40x^7y^3$.

Factor the polynomials, including the coefficients:

$$2^3x^7y \quad \text{and} \quad 2 \cdot 5xy^3$$

The LCM will include factors of 2, 5, x, and y. The greatest power of 2 that appears anywhere is 2^3; of 5, 5; of x, x^7; of y, y^3. So the LCM is $2^3 \cdot 5x^7y^3 = 40x^7y^3$.

b. The LCM of $(3m - 2n)(m + 5)$ and $(2n - 3m)(m + n)$ is $(3m - 2n)(m + 5)(m + n)$.

The polynomials are already in factored form. Note that $3m - 2n$ in the first polynomial and $2n - 3m$ in the second differ by a factor of -1, which doesn't count in prime factorizations and LCM. In short, only one of $3m - 2n$ and $2n - 3m$ counts. The LCM, therefore, can be written as $(3m - 2n)(m + 5)(m + n)$.

c. The LCM of $14(a^2 - b^2)$, $21(a^2 + 2ab + b^2)$, and $12(a + b)$ is $84(a - b)(a + b)^2$.

To find the LCM, factor each of the polynomials:

$$14(a^2 - b^2) = 2 \cdot 7(a - b)(a + b)$$
$$21(a^2 + 2ab + b^2) = 3 \cdot 7(a + b)^2$$
$$12(a + b) = 2^2 \cdot 3(a + b)$$

So the LCM will include 2 to the second power, $a + b$ to the second power, 3, 7, $a - b$, and $a + b$ to the second power. The LCM is $2^2 \cdot 3 \cdot 7(a - b)(a + b)^2 = 84(a - b)(a + b)^2$.

ADDING AND SUBTRACTING USING THE LCM

When adding and subtracting rational expressions with different denominators, most people use the LCM .

Read through this first example to see how the LCM method works.

EXAMPLE: Adding, LCM

Find $\dfrac{x}{(x+1)(x-4)} + \dfrac{x}{(x-4)^2}$.

SOLUTION

To add these two fractions, we need to convert them to equivalent fractions with a common denominator; we'll use the LCM of the denominators. The LCM is $(x+1)(x-4)^2$.

    To convert the first fraction, note that the denominator differs from the LCM by a factor of $x - 4$. So convert to an equivalent fraction whose denominator is the LCM by multiplying top and bottom by $(x - 4)$:

$$\frac{x}{(x+1)(x-4)} = \frac{x(x-4)}{(x+1)(x-4)^2}$$

The denominator of the second fraction differs from the LCM by a factor of $x + 1$. Convert the second fraction to an equivalent fraction whose denominator is the LCM by multiplying top and bottom by $x + 1$.

$$\frac{x}{(x-4)^2} = \frac{x(x+1)}{(x-4)^2(x+1)}$$

Now add:

$$\frac{x}{(x+1)(x-4)} + \frac{x}{(x-4)^2} = \frac{x(x-4)}{(x+1)(x-4)^2} + \frac{x(x+1)}{(x-4)^2(x+1)}$$

$$= \frac{x^2 - 4x + x^2 + x}{(x+1)(x-4)^2}$$

$$= \frac{2x^2 - 3x}{(x+1)(x-4)^2}$$

Try the next example on your own before reading the solution.

EXAMPLE: Subtracting, LCM

Find $\dfrac{1}{x^2 + x - 12} - \dfrac{1}{x^2 - 7x + 12}$.

SOLUTION

First, find the LCM of the denominators. To do that, factor them:

$$x^2 + x - 12 = (x + 4)(x - 3)$$

$$x^2 - 7x + 12 = (x - 4)(x - 3)$$

So the LCM is $(x - 3)(x - 4)(x + 4)$.

The first fraction needs an extra factor of $x - 4$ in the denominator to make the denominator the LCM:

$$\frac{1}{x^2 + x - 12} = \frac{1}{(x + 4)(x - 3)} = \frac{1(x - 4)}{(x - 3)(x - 4)(x + 4)}$$

The second fraction needs an extra factor of $x + 4$ to make the denominator into the LCM:

$$\frac{1}{x^2 - 7x + 12} = \frac{1}{(x - 4)(x - 3)} = \frac{1(x - 4)}{(x - 3)(x - 4)(x + 4)}$$

Subtract:

$$\frac{1}{x^2 + x - 12} - \frac{1}{x^2 - 7x + 12}$$

$$= \frac{x - 4}{(x - 3)(x - 4)(x + 4)} - \frac{x + 4}{(x - 3)(x - 4)(x + 4)}$$

$$= \frac{(x - 4) - (x + 4)}{(x - 3)(x - 4)(x + 4)}$$

$$= \frac{x - 4 - x - 4}{(x - 3)(x - 4)(x + 4)}$$

$$= \frac{-8}{(x - 3)(x - 4)(x + 4)}$$

Summary

Rational expressions

A *rational expression* is a glorified fraction, the quotient of two polynomials. A rational expression is not defined whenever its denominator is 0. That is, roots of the denominator are not allowed as variable values.

Simplified form

A rational expression is in *simplified* form if its numerator and denominator have no common polynomial factors. Reduce to simplified form by canceling common factors:

$$\frac{PR}{QR} = \frac{P\cancel{R}}{Q\cancel{R}} = \frac{P}{Q}$$

where P is any polynomial and Q and R are nonzero polynomials.

Multiplying rational expressions

As with fractions, multiply numerators and multiply denominators. That is,

$$\frac{P}{Q} \cdot \frac{R}{S} = \frac{PR}{QS}$$

where P and R are any polynomials and Q and S are nonzero polynomials.

Dividing rational expressions

As with fractions, flip the second expression and multiply. That is,

$$\frac{P}{Q} \div \frac{R}{S} = \frac{PS}{QR}$$

where P is any polynomial and Q, R, and S are nonzero polynomials.

Adding and subtracting rational expressions

As with fractions, first convert to expressions with a common denominator and then add or subtract numerators. The common denominator may be the product of the two denominators, or it may be their LCM. In general,

$$\frac{P}{Q} \pm \frac{R}{S} = \frac{PS \pm RQ}{QS}$$

where P and R are any polynomials and Q and S are nonzero polynomials.

Sample Test Questions

Answers to these questions begin on page 415.

1. Which of the following are rational expressions?

   A. $\dfrac{3 + \sqrt{x}}{3 - \sqrt{x}}$

   B. $\dfrac{3 + x^3}{3 - x^3}$

   C. $\dfrac{3 + x^x}{3 - x^x}$

   D. $\dfrac{3x + x^{-3}}{3 - x^{-3}}$

2. Evaluate each rational expression for the given variable values.

   A. $\dfrac{4 + x}{x - 6}$ for $x = 8$

   B. $-\dfrac{2}{9}$ for $z = -3$

   C. $\dfrac{p + 1}{p^2 - 1}$ for $p = -1$

   D. $\dfrac{a^2 + b^2 + c^2}{a^2 + b^2 - c^2}$ for $(a, b, c) = (3, 2, 1)$

3. For each rational expression, say what values of the variable are allowed. Give your answer in set notation (see Chapter 8).

   A. $\dfrac{x - 3}{x + 4}$

   B. $\dfrac{x + 1}{6x}$

   C. $\dfrac{7}{x^2 + 3}$

   D. $\dfrac{x - 5}{x^2 - 6x + 5}$

4. Express each rational expression in simplified form.

   A. $\dfrac{3(x + 7)(x - 8)}{9(x - 8)(x + 6)}$

   B. $\dfrac{2x - 8}{(4 - x)(4 + x)}$

C. $\dfrac{-x-1}{x-1}$

D. $\dfrac{x^3-x}{6x^4+2x^3-8x^2}$

5. Find each product.

A. $\dfrac{3}{x+1} \cdot \dfrac{4}{x-1}$

B. $\dfrac{x-7}{3x+2} \cdot \dfrac{3x+2}{x-9}$

C. $\dfrac{x^2-4}{3x-1} \cdot \dfrac{-3x-1}{x^2+4x+4}$

D. $\dfrac{x-4}{2x^2+5x-3} \cdot \dfrac{2x-1}{x^2-x-12}$

6. Find each quotient.

A. $\dfrac{2}{x^2+1} \div \dfrac{6}{x-1}$

B. $(x^2-9) \div \dfrac{x^2-6x+9}{x+3}$

C. $\dfrac{x^2+3x-18}{2x^2-9x+10} \div \dfrac{x^2-9x+18}{2x-5}$

7. Find the LCM of each pair of polynomials.

A. $45x^2(x+1)$ and $6x^3(x-1)$

B. $2(x+8)$ and $6(x-9)$

C. x^2+x-6 and $x^2+5x-14$

8. Add or subtract, as specified.

A. $\dfrac{2+x}{x-1} + \dfrac{3-x}{1-x}$

B. $\dfrac{x+1}{x+3} - \dfrac{x+2}{x+4}$

C. $\dfrac{1+x}{1-x} - \dfrac{1-x}{1+x}$

D. $\dfrac{x+1}{x^2-3x} + \dfrac{x+1}{x^2+2x}$

E. $\dfrac{x^2+7x+1}{6x^2-x-7} - \dfrac{7x+2}{6x^2-x-7}$

Answers to Sample Test Questions

Chapter 1 Answers
Pages 48–49

1.
   A. 22
   B. 38

2.
   A. 50
   B. 8

3.
   A. $-3a - 14$
   B. $15x - 6y + 3$
   C. $2m - 1$
   D. $-\frac{1}{3}p + q$

4. There are $\frac{2}{3}\left(\frac{1}{2}a + 7\right)$ apples in the bag when Ethel and Thelma get home.

5.
   A. No
   B. Yes

6.
   A. $x = 29$
   B. $x = -10$
   C. $x = 5$

7.

    A. $y = 18$

    B. $y = 95$

    C. $x = -\dfrac{8}{3}$

8.

    A. $x = 19$

    B. $x = -\dfrac{5}{2}$

    C. $x = \dfrac{19}{10}$

9.

    A. all real numbers

    B. $x = -7$

    C. $x = 6$

    D. no solution

    E. $x = -\dfrac{7}{3}$

10. David has shrunk 21 articles of clothing in the laundry.

11. Gerald can afford to spend 9 hours working at the coffee shop.

12. There were 4 apples in Ethel and Thelma's bag at the beginning of their journey.

Chapter 2 Answers

1.

A.

B.

all real numbers

2.

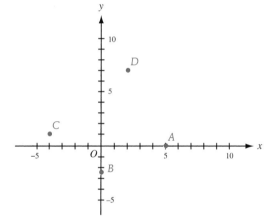

C and D are in the upper half-plane. None of the points is in quadrant III.

3.

A. $\dfrac{5}{2}$

B. $\sqrt{17}$

C. $\sqrt{72} = 6\sqrt{2}$

4.

A. $-\dfrac{1}{4}$

B. 1

C. $\dfrac{1}{2}$

D. undefined

5. (2, 3)

387

6.

    A. not linear

    B. linear

    C. linear

    D. linear

7.

x	y
−5	0
−2	1
1	2

(Table values may vary.)

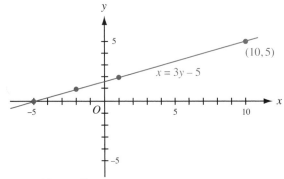

When $x = 10$, $y = 5$.

8.

    A.

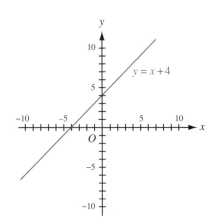

B.

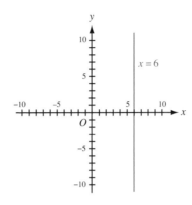

C.

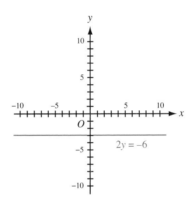

D.

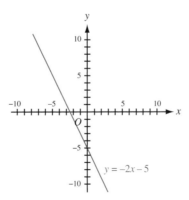

E.

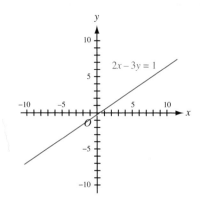

9.

   A. $y = 2$

   B. $y = -\dfrac{4}{3}x$

   C. $y = -x - 3$

   D. $x = -\dfrac{7}{2}$

   E. $y = 3x - 2$

10. $a = 15$

11. $y - 1 = -2(x - 3)$, or $2x + y = 7$

12. $5x - 4y = -2$

13. $y = 0$

14. No

15.

   A. Helmut recruited volunteers at an average rate of 3.5 volunteers per hour.

   B. $n = 3.5t + 4$

   C. Helmut can expect to have $3.5(7.5) + 4 = 30.25$, or just over 30, volunteers by 7 P.M.

16.

  A. In July, Elizabeth's hair was 7.5 inches long. In November, her hair was 12.3 inches long.

  B. At $t = 18.75$, which corresponds to about February of 2007.

  C.

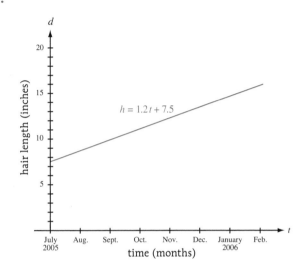

17.

  A. After 15 minutes, the bus is 14 miles from Disney World.

  B. It takes the bus 6 minutes to travel 4 miles, to the point $d = 20$.

  C. The d-intercept is 24 miles; that's where the bus starts out at time $t = 0$, the distance of the hotel from Disney World.

  D. The slope is $-\dfrac{2}{3}$ miles per minute; that's the speed of the bus (negative because the distance from Disney World is decreasing).

  E. $d = -\dfrac{2}{3}t + 24$

  F. The trip will take 36 minutes.

Chapter 3 Answers

1.

   A. (–3, 9) is not a solution to the simultaneous equations (though it is a solution to the first).

   B. (1, –2) is a solution to the simultaneous equations.

   C. $\left(0, \frac{5}{7}\right)$ is not a solution to the simultaneous equations (though it is a solution to the second).

2. The unique solution is (2, 1).

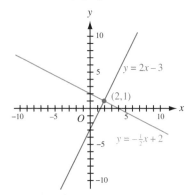

3. This system has no solutions.

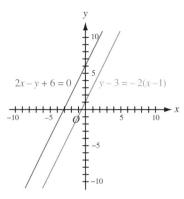

4. The unique solution is (2, 3).

5. The solutions are all the ordered pairs of the form $(x, y) = (-9 + 4y, y)$, where y can be any real number.

6. The unique solution is (–2, 4).

7. The unique solution is $\left(\dfrac{1}{2}, \dfrac{2}{3}\right)$.

8. The unique solution is (–5, –6).

9. The contest starts at 10 A.M. and lasts 3 hours.

10. 37 quarters and 53 dimes.

Chapter 4 Answers

1.
  - A. $\dfrac{1}{25}$
  - B. $-\dfrac{1}{343}$
  - C. 64
  - D. $\dfrac{625}{81}$

2.
  - A. $\dfrac{1}{n^2 m^4}$
  - B. $\dfrac{1}{x^8}$
  - C. $\dfrac{y^2}{2}$
  - D. 9

3.
  - A. $\sqrt{-36}$ is not a real number
  - B. $\dfrac{11}{7}$
  - C. $\sqrt{-(13)^2}$ is not a real number
  - D. 13

4.
  - A. −6
  - B. $\sqrt[4]{-25^2}$ is not a real number
  - C. $\dfrac{3}{2}$
  - D. −2

5.

    A. $5\sqrt{3}$

    B. $2\sqrt{15}$

    C. $4y\sqrt{6x}$

    D. $12x\sqrt{x}$

6.

    A. $2\sqrt[3]{15}$

    B. $x \cdot \sqrt[4]{x^3}$

    C. $-4x^2$

    D. $2y \cdot \sqrt[5]{8y^2}$

7.

    A. $2\sqrt{15}$

    B. 24

    C. $5pq\sqrt{3p}$

    D. $\dfrac{2\sqrt{2y}}{x}$

8.

    A. $3\sqrt{2}$

    B. $\sqrt{6} + \sqrt{10}$

    C. $4\sqrt{3}$

    D. $2\sqrt{10}$

9.

    A. $\dfrac{4}{7}\sqrt{7}$

    B. $\dfrac{-2\sqrt{15}}{5}$

    C. $\dfrac{\sqrt{6} + \sqrt{2}}{2}$

    D. $\dfrac{\sqrt[3]{4} + 2}{2}$

Chapter 5 Answers

Pages 224–227

1.

    A. polynomial of degree 2

    B. not a polynomial

    C. polynomial of degree 0

    D. polynomial of degree 5

    E. not a polynomial

    F. polynomial in two variables of degree 3

    G. polynomial in three variables of degree 1

2.

    A. linear binomial

    B. quintic monomial

    C. cubic trinomial

    D. quartic binomial

3.

    A. $x^2 + x + 7$

    B. $2x^2 + 6x - 9$

    C. $2x^2 + x + 7$

    D. $-3x^3 + x^2 + 4x - 11$

4.

    A. $2x^3 - 8x^2 + 14x$

    B. $-x^3y - xy^3 + x^2y + xy^2$

    C. $x^2 + 8x + 15$

    D. $x^2 + 4x - 12$

    E. $6x^2 - 19x + 10$

    F. $-3x^2 + xy + 14y^2$

    G. $x^4 + 2x^3 - 17x^2 + 2x + 12$

5. quotient $2x^2 - 3x + 6$, remainder -16

$$
\begin{array}{r}
2x^2 - 3x + 6 \\
x + 5 \overline{)\ 2x^3 + 7x^2 - 9x + 14} \\
\underline{-(2x^3 + 10x^2)} \\
-3x^2 - 9x + 14 \\
\underline{-(-3x^2 - 15x)} \\
6x + 14 \\
\underline{-(6x + 30)} \\
-16
\end{array}
$$

6. quotient $x^2 + 4x + 9$, remainder 20

$$
\begin{array}{r|rrrr}
4 & 1 & 0 & -7 & -16 \\
  &   & 4 & 16 & 36 \\
\hline
  & 1 & 4 & 9 & 20
\end{array}
$$

7.

   A. $4(3x - 2)$

   B. $z^2(z^5 - z^3 + 1)$

   C. $-2xy(x^2 - 5xy + 3y^2)$

8.

   A. $(y - 4)(y^2 + 2)$

   B. $(2x - 1)(y + 3)$

   C. $(x + 7)(4x + 9)$

   D. $3x(x - y)(y - 2)$

   E. $(a - 2)(b + 3c + 7d)$

9.

   A. $(x + 6)(x - 6)$

   B. $(10 + y)(10 - y)$

   C. $(x - 8)^2$

   D. $(3x + 1)^2$

   E. $3(1 - 3x)(1 + 3x)$

    F. $-2(z-11)^2$

    G. $(5x+3)^2$

    H. $(x^2-7)^2$

10.

    A. $(x+1)(x+7)$

    B. $(x-2)(x-5)$

    C. $(x-6)(x+1)$

    D. $(x+3)(x-2)$

    E. $3(x+6)(x-2)$

11.

    A. $(x+1)(2x-1)$

    B. $(x+4)(3x+1)$

    C. $(x-3)(2x+7)$

    D. $(x-5)(3x-2)$

    E. $(2x+1)(4x-3)$

12.

    A. $x^4-81=(x^2+9)(x^2-9)=(x^2+9)(x+3)(x-3)$

    B. $(x-1)^2-4=((x-1)+2)((x-1)-2)=(x+1)(x-3)$

    C. $y^2-6y+9-25z^2=(y-3)^2-(5z)^2=(y+5z-3)(y-5z-3)$

    D. does not factor

    E. $x^3+x^2-x-1=(x+1)(x^2-1)=(x-1)(x+1)^2$

    F. does not factor

    G. $(x^4+4+4x^2)-4x^2=(x^2+2)^2-(2x)^2$
       $=(x^2+2x+2)(x^2-2x+2)$

13. $a(21-a)=80$, or $a^2-21a+80=0$

14.

    A. Emma fetches a total of xy sticks.

    B. Ruby fetches a total of $xy-x+6y-6$ sticks.

    C. $6y-x-6=3$

    D. Ruby fetches 15 sticks per hour.

Chapter 6 Answers

Pages 262–264

1. $-3, -1, \frac{1}{2},$ and 1

2.

    A. not quadratic (linear)

    B. not quadratic (variable in denominator)

    C. quadratic

    D. not quadratic (variable in radicand)

    E. not quadratic (quintic)

3.

    A. $x = \dfrac{2}{\sqrt{5}}$ or $x = -\dfrac{2}{\sqrt{5}}$

    B. no solution

    C. $x = 2\sqrt{7}$ or $x = -2\sqrt{7}$

    D. $x = \dfrac{3}{5}\sqrt{15}$ or $x = -\dfrac{3}{5}\sqrt{15}$

4.

    A. $x = 5$ or $x = 1$

    B. $x = -5$ only

    C. $x = 1 + \sqrt{5}$ or $x = 1 - \sqrt{5}$

    D. $x = 2$ or $x = -5$

5.

    A. -3 and -8

    B. 0 and 7

    C. $\dfrac{9}{2}$ and $-\dfrac{5}{3}$

    D. $-\dfrac{3}{4}$ and $\dfrac{5}{6}$

6.

    A. $x = 2$ or $x = 5$

    B. $y = 9$ or $y = -2$

   C. $x = 0$ or $x = -\dfrac{5}{2}$

   D. $x = -\dfrac{7}{2}$ only

   E. $x = -8$ or $x = \dfrac{2}{3}$

   F. $t = -3$ or $t = 8$

7.

   A. $x^2 + \underline{10x} + 25$

   B. $z^2 - \underline{18z} + 81$

   C. $x^2 - \underline{7x} + \dfrac{49}{4}$

   D. $16x^2 + \underline{24x} + 9$

8.

   A. 64

   B. 169

   C. $\dfrac{25}{4}$

   D. $\dfrac{1}{36}$

9.

   A. $y = 12$ or $y = -2$

   B. no solution

   C. $x = 1 + \sqrt{6}$ or $x = 1 - \sqrt{6}$

   D. $x = 7$ or $x = -2$

10.

   A. discriminant is 49: two real roots

   B. discriminant is 0: one real root

   C. discriminant is -36: no real roots

   D. discriminant is 24: two real roots

11.

A. $x = 3$ or $x = -\dfrac{1}{2}$

B. $x = \dfrac{3}{5}$ only

C. no solution

D. $x = -4 + \sqrt{6}$ or $x = -4 - \sqrt{6}$

The number of solutions to each equation is the same as the number of roots of the corresponding polynomial, as predicted in question 10.

12. Ella and Nina's room is 15 feet long by 6 feet wide.

13.

A. After 2 seconds, the rock is 34 feet from the ground.

B. The rock is 25 feet up at 0.5 second and again at 2.375 seconds.

C. The rock spends 3 seconds in the air.

14. The three sides have lengths 8, 15, and 17.

Chapter 7 Answers

Pages 307–314

1.

A. $y = (x - 2)^2 + 5$, with $a = 1$, $k = 2$, and $h = 5$

B. $y = -\left(x + \dfrac{7}{2}\right)^2 + \dfrac{49}{4}$, with $a = -1$, $k = -\dfrac{7}{2}$, and $h = \dfrac{49}{4}$

C. $y = 2x^2 + 8$, with $a = 2$, $k = 0$, and $h = 8$

D. $y = \dfrac{1}{3}(x + 4)^2 - \dfrac{22}{3}$, with $a = \dfrac{1}{3}$, $k = -4$, and $h = -\dfrac{22}{3}$

E. $y = -4(x - 1)^2$, with $a = -4$, $k = 1$, and $h = 0$

2.

A. vertex at $(-7, 9)$

B. vertex at $(-1, -8)$

C. vertex at $\left(\dfrac{15}{4}, -\dfrac{3}{8}\right)$

D. vertex at $(0, 5)$

3.

A. axis of symmetry is $x = 0$

B. axis of symmetry is $x = -5$

C. axis of symmetry is $x = \dfrac{6}{5}$

4. The roots are -4 and -1.

5.

A. twice

B. once

C. none

D. twice

E. twice

6.

A. x-intercepts: $\dfrac{1}{2}$ and 6; y-intercept: 6

B. x-intercepts: 2 and -4; y-intercept: -8

C. no *x*-intercepts; *y*-intercept: 9

D. *x*-intercept: $-\dfrac{1}{3}$; *y*-intercept: 1

E. no *x*-intercepts; *y*-intercept –53

7.

A.

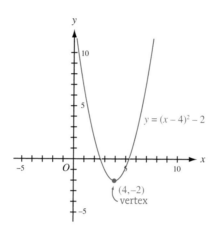

$y = (x - 4)^2 - 2$

$(4, -2)$ vertex

B.

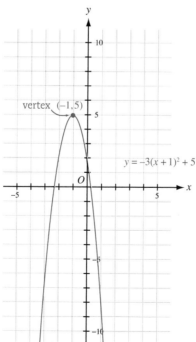

vertex $(-1, 5)$

$y = -3(x + 1)^2 + 5$

C.

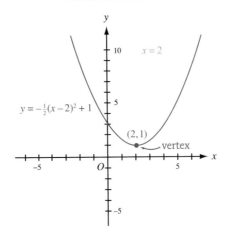

D.

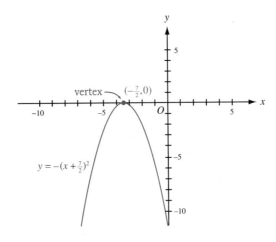

8.

A.

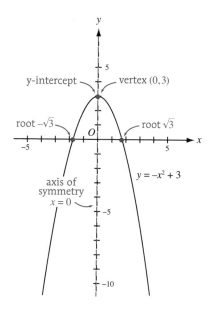

B.

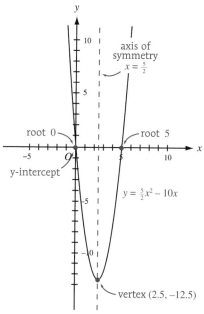

C.

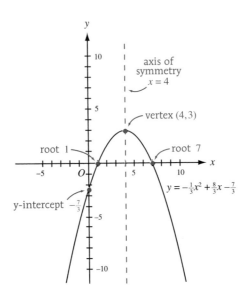

D.

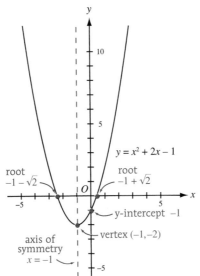

9.

    A. $y = (x - 2)^2$ or $y = x^2 - 4x + 4$

    B. $y = -(x + 1)(x - 2)$, or $y = -x^2 + x + 2$

    C. $y = 3x(x + 4)$, or $y = 3x^2 + 12x$

    D. $y = \frac{1}{4}x^2 + 3$

10.

    A. $y = -(x - 3)^2 + 2$

    B. $y = \frac{1}{10}(x + 5)^2$

11.

    A. $y = \frac{-1}{6}(x + 4)(x - 3)$

    B. $-(x + 2)^2$, or $-x^2 - 4x - 4$

12. *a* is negative (The vertex of this parabola is below the *x*-axis; if *a* were positive, the parabola would open up and cross the *x*-axis.)

13.

    A.

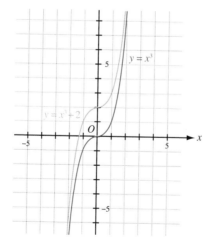

B.

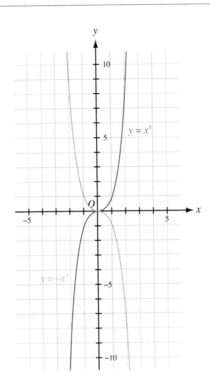

C.

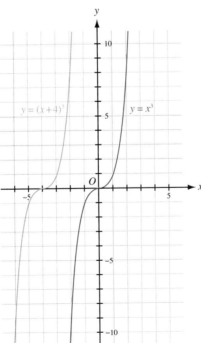

14.

x	y
24	–3
15	–2
8	–1
3	0
0	1
–1	2
0	3
3	4
8	5

Chapter 8 Answers

1.

    A. $x > 3$

    B. $x \leq -0.5$

    C. $-1 \leq x < 4$

2.

    A. $[2, 7]$

    B. $(-2, -1)$

    C. $[-3, \infty)$

3.

    A.

$x < 2$

    B.

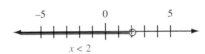

$x \geq -4$

    C.

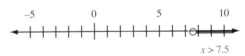

$x > 7.5$

    D.

$-3 \leq x \leq 3$

    E.

$-1 < x \leq 2$

    F.

$-3.5 < x < 1.5$

4.

A.

[0, 2)

B.

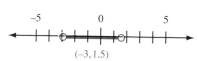

(−3, 1.5)

C.

[0.25, 1]

D.

(−∞, 5)

5.

A. $\{x : x \le 4\}$, or $(-\infty, 4]$

B. $\{x : x > 60\}$, or $(60, \infty)$

C. $\{x : -7 \le x \le -3\}$, or $[-7, -3]$

D. $\{x : x < -\dfrac{3}{2}\}$, or $(-\infty, -\dfrac{3}{2})$

E. $\{x : 1 \le x < 3\}$, or $[1, 3)$

F. $\{x : x \ge 2\}$, or $[2, \infty)$

6.

A. $\{x : x < 1\}$

$x < 1$

B. $\{x : x > 6\}$

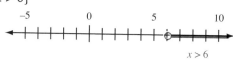

$x > 6$

C. $\{x : x \le \frac{1}{2}\}$

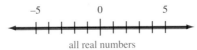

$x \le \frac{1}{2}$

D. all real numbers

all real numbers

7.

A. $\{x : 1 < x < 3\}$, or $(1, 3)$

B. no solution

C. $\{x : x \ge 2\}$, or $[2, \infty)$

D. $\{3\}$

8.

A. $c < \frac{1}{2}$

B. $b \ge 4$

C. $12 \le g$

D. $3 \le p \le 7$

E. $8h > 70$

9.

A. $21x < 2 \cdot 91$ (Convert 1 hour and 31 minutes to 91 minutes.)

B. $x < \frac{26}{3}$

C. David will be able to see at most 8 episodes of *Arrested Development*.

10. The third side s satisfies $13 - 8 < s < 13 + 8$, or $5 < s < 21$.

11.

A.

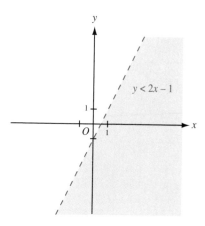

B.

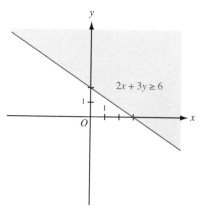

12.

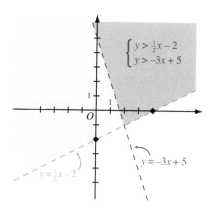

13.

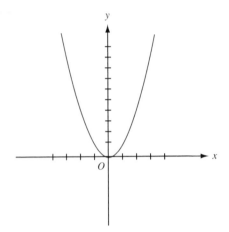

14.

   a.

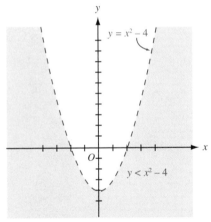

   b. $\{x : x < -2$ OR $x > 2\}$ (These are the shaded points along the x-axis on the graph of $y < x^2 - 4$.)

Chapter 9 Answers

1.

    A. not rational (because of $\sqrt{x}$)

    B. rational expression

    C. not rational (because of x^x)

    D. rational expression (equivalent to $\dfrac{3x^3+1}{3x^3-1}$)

2.

    A. 6

    B. $-\dfrac{2}{9}$

    C. undefined

    D. $\dfrac{7}{6}$

3.

    A. $\{x : x \neq -4\}$

    B. $\{x : x \neq 0\}$

    C. all real numbers

    D. $\{x : x \neq 5, x \neq 1\}$

4.

    A. $\dfrac{x+7}{3(x+6)}$

    B. $\dfrac{-2}{x+4}$

    C. $\dfrac{-x-1}{x-1}$

    D. $\dfrac{x+1}{2x(3x+4)}$

5.

A. $\dfrac{12}{x^2 - 1}$

B. $\dfrac{x - 7}{x - 9}$

C. $\dfrac{-(x - 2)(3x + 1)}{(3x - 1)(x + 2)}$

D. $\dfrac{1}{(x + 3)^2}$

6.

A. $\dfrac{x - 1}{3(x^2 + 1)}$

B. $\dfrac{(x + 3)^2}{x - 3}$

C. $\dfrac{x + 6}{(x - 2)(x - 6)}$

7.

A. $90x^3(x + 1)(x - 1)$

B. $6(x + 8)(x - 9)$

C. $(x - 2)(x + 3)(x + 7)$

8.

A. $\dfrac{2x - 1}{x - 1}$

B. $\dfrac{-2}{x^2 + 7x + 12}$

C. $\dfrac{4x}{1 - x^2}$

D. $\dfrac{(x + 1)(2x - 1)}{x(x - 3)(x + 2)}$

E. $\dfrac{x - 1}{6x - 7}$

Glossary

A

additive inverse of a real number the negative of the number; the number that, when added to the original number, gives 0; for example, –7 is the additive inverse of 7, and $\frac{1}{2}$ is the additive inverse of $-\frac{1}{2}$

algebraically using algebra—variables, numbers, operations, equations—especially as contrasted with words or graphs

algebraic expression one or more numbers and variables combined together with addition, subtraction multiplication, or division

arithmetic expression one or more real numbers combined together using only addition, subtraction, multiplication, and division

axis one of two number lines (usually, the x-axis and the y-axis) that help identify points in the coordinate plane

axis of symmetry of a parabola the imaginary vertical line that divides a parabola into two mirror-image halves; the axis of symmetry is the line $x = k$ for the parabola $y = a(x - k)^2 + h$, or equivalently, the line $x = -\frac{b}{2a}$ for the parabola $y = ax^2 + bx + x$

B

base the number being raised to a power; the a in the exponential expression a^n

binomial a polynomial with two terms, like $-3x^3 + 7$

bounded interval an interval that has two endpoints and a definite length, like $[1, 3)$ or $(-2, 0)$

C

canceling reducing a fractional expression denoted by crossing out the common factors, as in $\dfrac{3}{12} = \dfrac{\cancel{3}}{\cancel{12}_4}$

Cartesian plane another name for the coordinate plane, named after the seventeenth-century mathematician René Descartes

closed interval an interval that includes its endpoints; for example, [–3, 0] or [7, ∞)

coefficient the number part of a term that involves variables; for example, the coefficient of –2*xy* is –2

collinear points three or more points that lie on the same straight line

common multiple of two integers a number that is divisible by both; for example, 80 is a common multiple of 4 and 10

common multiple of two polynomials a polynomial that is divisible by both; for example, $x^2 - 1$ is a common multiple of $x + 1$ and $x - 1$

completing the square expressing a quadratic equation (such as $ax^2 + bx + c = 0$) in the form $a(x + p)^2 = d$, with the ultimate goal of solving the equation

compound inequality two inequality statements in one, as in $3 < x - 7 < 5$

consistent system of equations a system of equations with at least one solution

constant a quantity that does not change in a particular context, like the number 32 in $F = \dfrac{9}{5}C + 32$ or the slope *m* in $y = mx + b$

constant polynomial a polynomial of degree 0, like the polynomial –9

constant term a number-only term within any mathematical expression; for example, the constant term in $3\sqrt{x} + x^2 - 4$ is –4

coordinate either of the two numbers in an ordered pair

coordinate plane a system for identifying points in the plane consisting of two perpendicular number lines (usually, the *x*- and *y*-axes) intersecting at a point (the origin); a point is identified by an ordered pair that gives its distance from the two number lines

cross-term in a trinomial of the form $ax^2 + bxy + c$, the middle term bxy, which that involves both xs and ys

cube of a number the third power of the number; for example, the cube of 5 is 5^3, or 125

cube root of a number the number whose cube is the original number, denoted with $\sqrt[3]{\ }$; for example, $\sqrt[3]{8} = 2$ and $\sqrt[3]{-125} = -5$

cubic polynomial a one-variable polynomial of degree 3, like $2x^3 + x^2 - 7$

D

degree of a monomial the sum of the exponents of all the variables in the monomial; for example, the degree of $7xyz^2$ is 4

degree of a polynomial the highest degree of any monomial terms within a polynomial; for example, the degree of $7xy^2 - xy^5 + 7$ is 6

dependent linear equations in two variables two or more equations that describe lines that have the same slope, so the lines are parallel or identical

difference of squares an expression in the form $a^2 - b^2$, which factors as $(a + b)(a - b)$

discriminant of a quadratic a number, given by $b^2 - 4ac$ for the polynomial $ax^2 + bx + c$, whose sign determines how many real roots the polynomial has: if the discriminant is positive, the polynomial has two real roots; if negative, no real roots; if zero, one real root

distributive property of multiplication over addition property of real numbers that says that $a(b + c) = ab + ac$ for real a, b, c

dummy variable a variable whose particular name doesn't matter; for example, the x in the set $\{x : x > 2\}$ is a dummy variable since $\{y : y > 2\}$ denotes the exact same collection of numbers

E

empty set the set with no elements, denoted by {} or by $\emptyset$

endpoint of an interval a boundary point of the interval; for example, 2 is an endpoint of the interval (2, 9), and –4 is an endpoint of the interval $(-\infty, -4]$

equal sign the sign =, which relates two expressions that identify the same point on the number line, like 2 + 2 and 4 in the equation 2 + 2 = 4

equation a mathematical statement that two expressions have the same value

equivalent equations equations that have the same solutions

equivalent fractions fractions that represent the same quantity, like $\frac{4}{6}$ and $\frac{-20}{-30}$

equivalent inequalities two or more inequalities that have the same solutions; for example, $-2x < 8$ and $x > -4$ are equivalent inequalities

equivalent rational expressions two or more rational expressions that give the same value whenever the same variable values are plugged in; for example, $\frac{x}{x(x-1)}$ and $\frac{1}{x-1}$ are equivalent rational expressions unless $x = 0$

evaluate an expression simplify the expression until you get a real number

exponent the number of copies of the base being multiplied together; the n in the expression a^n

exponential notation a common way of writing repeated multiplication where a^n means n number of as multiplied together; for example, $2^3 = 2 \cdot 2 \cdot 2$

exponentiation raising a number to a power; repeated multiplication

F

factor a polynomial express the polynomial as products of polynomials of smaller degree; for example, $x^2 - 4xy - 5y^2$ factors as $(x - 5y)(x + y)$

fourth root of a number a number whose fourth power is the original number, denoted with $\sqrt[4]{}$; for example, $\sqrt[4]{16} = 2$

G

graph an equation draw the graph of the equation, on the number line if the equation involves one variable, on the coordinate plane if it involves two

graph of a one-variable equation the set of all points x on the number line that are solutions to the equation

graph of a two-variable equation the set of all points (x, y) in the coordinate plane that are solutions to the equation

greater than or equal to sign ≥; write $a \geq b$ if a is to the right of or at b along the number line

greater than sign >; write $a > b$ if a is to the right of b along the number line

H

half-closed interval another name for half-open interval; a (bounded) interval that includes one endpoint but not the other, like $[1, 3)$

half-open interval a (bounded) interval that includes one endpoint but not the other; for example, $(2, 5]$ or $[0, 1)$

horizontal shift of a graph a graph obtained by replacing x by $x - k$ in the graph equation; for example, the graph of $y = (x - 3)^2$ is a horizontal shift of the graph of $y = x^2$ to the right by 3 units, and the graph of $y = -4(x + 1)^2$ is a horizontal shift of the graph of $y = -4x^2$ to the left by 1 unit

horizontal translation of a graph another name for horizontal shift; for example, the graph of $y = \frac{1}{3}(x - 4)^2 + 7$ is a horizontal translation of the graph of $y = \frac{1}{3}x^2 + 7$ to the right by 4 units

hypotenuse the longest side of a right triangle

I

in terms of relying on, as a variable; for example, you can express the temperature
of something in degrees Fahrenheit in terms of its temperature in degrees
Celsius (C), as $\frac{9}{5}C + 32$

inconsistent system of equations a system of equations that has no solutions

independent linear equations in two variables two equations whose graphs
describe intersecting lines

index of a root the n in an nth root expression like $\sqrt[n]{a}$

inequality a mathematical statement saying two quantities are not equal, like
$2 + 2 \neq 5$ or $x + 3 < 8$

inequality sign any of the signs $<$, $>$, $\neq$, $\leq$, and $\geq$, which compare two unequal
quantities

infinity a conceptual aid that represents roughly something greater than any real
number, denoted by the symbol ∞

intersection of two graphs the set of points common to both graphs

interval a continuous segment of the number line; for example, the set
$\{x : 1 < x < 2\}$ or the set $\{x : x \geq 9\}$, denoted in interval notation by $(1, 2)$ and
$[9, \infty)$, respectively

L

GLOSSARY

LCM least common multiple; for example, 24 is the LCM of 6 and 8, and
$x(x + 1)(x - 1)$ is the LCM of $x(x + 1)$ and $x(x - 1)$

lead coefficient the coefficient of the leading term in a one-variable polynomial;
for example, the lead coefficient of $-x^3 + 3x + 2$ is -1

leading term the highest-degree term in a one-variable polynomial; for example
$3x^2$ is the leading term of $3x^2 - 7x + 9$

least common multiple of two integers the smallest of the common multiples of the two integers; for example, 20 is the least common multiple of 4 and 10

least common multiple of two polynomials the smallest-degree polynomial that is divisible by both; for example, $x^2(x + 1)$ is the least common multiple of x^2 and $x(x + 1)$

left half-plane the points with negative x-coordinate in the standard xy-plane; those points to the left of the y-axis

leg either of the two shorter sides of a right triangle; each touches the right angle

less than or equal to sign $\leq$; write $a \leq b$ if a is to the left of or at b along the number line

less than sign $<$; write $a < b$ if a is to the left of b along the number line.

like terms in an expression, the terms that include the same variables to the same powers, like $-3x^2y$ and $29x^2y$, or $3a$ and $-a$

linear equation in one variable an equation that can be simplified to the form $Ax + B = Cx + D$, where A, B, C, and D are real numbers; it may have one solution, no solutions, or all real numbers as solutions

linear equation in two variables an equation that can be written in the form $Ax + By = C$, where A, B, and C are real numbers; its graph is a straight line

linear polynomial a polynomial of degree 1, like $x + y - 8$

lower half-plane the points with negative y-coordinate in the standard xy-plane; those points below the x-axis

lowest terms the state of any fractional expression whose numerator and denominator have no common factors other than ± 1 ; for example, $\frac{1}{4}$ and $\frac{1}{x + 2}$ are in lowest terms, but $\frac{-7}{-28}$ and $\frac{x + 3}{x^2 + 5x + 6}$ are not; lowest terms is also known as simplified, or reduced, form

M

maximum y-value of a graph the greatest height (along the y-axis) reached by the graph, a value that may or may not exist; for example, the maximum y-value of $y = -x^2 + 8$ is 8, but $y = x^2$ has no maximum y-value

minimum y-value of a graph the least height (along the y-axis) reached by the graph, a value that may or may not exist; for example, the minimum y-value of $y = x^2 - 7$ is –7, but $y = -x^2$ has no minimum y-value

monic polynomial a polynomial with lead coefficient 1, like $x^2 - 4x + 4$

monomial any single product of variables or numbers or both; for example, $-3xyz^2$ or y or 45

multiplicative inverse of a number the number you have to multiply by to get 1; the multiplicative inverse of a nonzero real number a is $\frac{1}{a}$

N

negative infinity a conceptual aid that represents roughly something less than any real number, denoted by $-\infty$

not equal to sign $\neq$; write $a \neq b$ if a and b identify different points on the number line

n^{th} power of a number the product of n copies of the number; for example, the third power of 4 is $4 \cdot 4 \cdot 4$, or 64

n^{th} root of a number a number whose n^{th} power is the original number; for example, the fifth root of 243 is 3 because $3^5 = 243$

O

one-dimensional needing only one number to identify a particular of its points, like a line or the circumference of a circle

open interval an interval that does not include its endpoints; for example, $(-2, -1)$ or $(-\infty, 3)$

order of operations the conventional hierarchy of arithmetic operations: first expand parentheses, then expand exponential expressions, then do multiplication and division, then do addition and subtraction

ordered pair a pair of numbers, like $(3, -7)$, that represents a point (x, y) in the coordinate plane or keeps track of values of the variables in a two-variable expression or equation, alphabetically ordered

origin the point $(0, 0)$ in the coordinate plane, where the two axes intersect

P

parabola the U-shape of the graphs of quadratic equations like $y = x^2$ or $y = ax^2 + bx + c$

parallel lines two or more lines that run in the same direction, never intersecting

perfect cube a number that is the cube of an integer, like 1, 125, and –8

perfect square a number that is the square of an integer, like 9, 64, and 0

perpendicular lines two lines that intersect at right angles

plug in a value for a variable in an expression replace every instance of the variable with the value; for example, plugging $x = 4$ into $3x^2 - 5x$ gives $3(4)^2 - 5(4)$, which can be evaluated to a real number

point-slope form the $y - y_0 = m(x - x_0)$ form of a linear equation, where m is the slope and (x_0, y_0) is any particular solution

polynomial any sum or difference of products of variables and numbers

power of a number the product of a number with itself several times; for example, 32 is a power of 2 because $32 = 2 \cdot 2 \cdot 2 \cdot 2 \cdot 2$

principal square root the positive square root of a number, denoted with the $\sqrt{\ }$ sign; for example, 25 has two square roots, 5 and –5, but only 5 is the principal square root

Q

quadrant one of four quarters of the coordinate plane created by the intersecting axes

quadrant I the set of points with positive x- and y-coordinates in the standard xy-plane

quadrant II the set of points with negative x- and positive y-coordinates in the standard xy-plane

quadrant III the set of points with negative x- and y-coordinates in the standard xy-plane

quadrant IV the set of points with positive x- and negative y-coordinates in the standard xy-plane

quadratic equation an equation that can be expressed in the form $ax^2 + bx + c = 0$, where $a \neq 0$; for example, $3x^2 - 4x = 7$ or $x^2 - 8x = 0$

quadratic formula a formula for the solutions to a quadratic equation in terms of the coefficients: the solutions to $ax^2 + bx + c = 0$ are given by
$$x = \frac{-b \pm \sqrt{b^2 - 4ac}}{2a}$$

quadratic polynomial a one-variable polynomial of degree 2, like $-x^2 - 3x + 4$

quartic polynomial a one-variable polynomial of degree 4, like $x^4 - 4x^2 + 4$

quintic polynomial a one-variable polynomial of degree 5, like $2x^5 - 4x^3 + 9$

R

radical a root expression written with a radical sign; for example, $\sqrt{2}$ and $\sqrt[3]{-7x}$

radical sign the sign $\sqrt{}$, which denotes principal square root, as in $\sqrt{25} = 5$, or any higher-degree root, as in $\sqrt[3]{8} = 2$

radicand an expression under a radical sign, like $x + 3$ in the expression $2\sqrt{x + 3}$

rational expression a quotient of two polynomials, like $\dfrac{x^2 + y}{y - 1}$

rationalizing the denominator a procedure for eliminating radicals from the denominator of an expression: multiply top and bottom by the smallest factor that will leave the denominator rational; for example, multiply top and bottom of $\dfrac{1}{\sqrt{3}}$ by $\sqrt{3}$, multiply top and bottom of $\dfrac{1}{\sqrt[3]{4}}$ by $\sqrt[3]{2}$, and multiply top and bottom of $\dfrac{1}{\sqrt{5}-1}$ by $\sqrt{5}+1$

rational number a number that can be expressed as a quotient of two integers, like $-\dfrac{1}{2}$ or 6 or 8.9

ray half a line, a piece with one endpoint that extends to infinity in one direction, like $(8, \infty)$ or $(-\infty, 4]$; rays are unbounded intervals

reciprocal of a fraction the fraction flipped over, top to bottom; the reciprocal of $\dfrac{2}{3}$ is $\dfrac{3}{2}$ and the reciprocal of $\dfrac{1}{7}$ is 7

reduced form of a fractional expression a fraction or rational expression whose numerator and denominator have no common factors other than ± 1; for example, $\dfrac{3}{4}$ is in reduced form but $\dfrac{9}{12}$ is not, and $\dfrac{x}{x+1}$ is in reduced form but $\dfrac{x^2 - x}{x^2 - 1}$ is not; reduced form is also known as simplified form or lowest terms

reflection of a graph over the x-axis the mirror image of the graph flipped over the x-axis, obtained by multiplying the expression for y by -1; for example, the graph of $y = -x^2$ is the reflection of the graph of $y = x^2$ over the x-axis, and the graph of $y = 2x^2$ is the reflection of the graph of $y = -2x^2$ over the x-axis

right half-plane the points with positive x-coordinate in the standard xy-plane; those points to the right of the y-axis

right triangle a triangle with a right angle

root of a polynomial a value for the variable that makes the (one-variable) polynomial evaluate to zero; for example, 3 is a root of the polynomial $2x^2 - 18$

roster notation a method of writing down a set by listing all its elements explicitly; for example, the set $\{2, 3, 5\}$ is written in roster notation

S

set any collection of numbers or other items, denoted by enclosing in curly braces; for example, the set of all even numbers between 1 and 9 is written $\{n : n$ even and $1 < n < 9\}$ or $\{2, 4, 6, 8\}$

set-builder notation a method of writing down a set by describing its elements rather than listing them; for example, the set $\{p : p$ is prime and $p < 6\}$ is written in set-builder notation

simplified form of a fractional expression a fraction or rational expression whose numerator and denominator have no common factors except ± 1; for example, $\frac{-2}{3}$ is in simplified form, but $\frac{-10}{15}$ is not, and $\frac{1}{x}$ is in simplified form, but $\frac{x}{x^2}$ is not; simplified form is also known as reduced form or lowest terms

simplified form of an expression a technical term for an expression in which there are no parentheses and no uncombined like terms, and every fractional expression is in lowest terms; also, no radicals in the denominator and no repeated factors in the radicand

simplified n^{th} root a root whose radicand is not divisible by any perfect n^{th} power

simplify an expression convert an expression to an equivalent expression in simplified form

simultaneous equations a system of two or more equations considered together, so the variables take on the same values across all equations

slope-intercept form the $y = mx + b$ form of a linear equation in two variables; m is the slope and b is the y-intercept

slope of a line a measure of the line's steepness; if a line passes through (x_1, y_1) and (x_2, y_2), its slope given by $\dfrac{y_2 - y_1}{x_2 - x_1}$

solution to an equation a value, or set of values, for the variable, or variables, that makes the equation true; for example, $x = -3$ is a solution to the equation $x^2 = 9$

solution to an inequality a variable value that make the inequality true; for example, $x = 8$ is a solution to the inequality $2x - 9 > 5$

solve an equation to find all solutions to the equation

square of a number the product of the number with itself; the second power of the number; for example, the square of 3 is 3^2, or 9

square root of a number a number whose square is the original number; for example, –3 is a square root of 9

standard form of a linear equation in two variables an $Ax + By = C$ form; note that this form is not unique: for example, $x + 2y = 3$ and $2x + 4x = 6$ are two equivalent equations in standard form

substitute a value for a variable in an expression replace every instance of the variable with the value; for example, substituting $a = -1$ into $3\sqrt{7-a}+a$ gives $3\sqrt{7-(-1)}+(-1)$, which can be evaluated to a real number

synthetic division long division of polynomials by binomials of the form $x - a$

system of equations two or more equations considered simultaneously, so the variables take on the same values across all equations

T

term a multiplicative bit being added or subtracted within any mathematical expression; for example, the polynomial $4x^2 - 6x - 10$ has three terms

trinomial a polynomial with three terms, like $4x^2 - 8xy + 3y^2$

two-dimensional needing two numbers to identify a particular of its points, like a plane or the surface of a sphere

U

unbounded interval an interval that extends forever in one or both directions, like $(-\infty, 0)$ or $[1, \infty)$ or $(-\infty, \infty)$

undefined expression an expression that has no value, like $\dfrac{7}{0}$ or 0^0 or $\dfrac{x}{x}$ for $x = 0$

upper half-plane the points with positive y-coordinate in the standard xy-plane; those points above the x-axis

V

variable an unknown or unspecified quantity, one that can take on any of a number of values

vertex of a parabola the turning point of the U-shape of the parabola, located at (k, h) for the parabola $y = a(x - k)^2 + h$; equivalently, at $\left(\frac{-b}{2a},\ c - \frac{b^2}{4a}\right)$ for the parabola $y = ax^2 + bx + c$

vertical compression of a graph another name for vertical stretch by a factor less than 1; for example, $y = \frac{1}{5}x^2$, which is a vertical stretch of the graph of $y = x^2$ by a factor of $\frac{1}{5}$, is also a vertical compression of $y = x^2$ by a factor of 5

vertical shift of a graph a graph obtained by adding a constant to the expression for y; for example, the graph of $y = x^2 + 8$ is a vertical shift of the graph of $y = x^2$ up by 8 units, and the graph of $y = 2(x + 1)^2 - 3$ is a vertical shift of the graph of $y = 2(x + 1)^2$ down by 3 units

vertical stretch of a graph a graph obtained by multiplying the expression for y by a positive factor; for example, the graph of $y = 2x^2$ is a vertical stretch of the graph of $y = x^2$ by a factor of 2, and the graph of $y = -\frac{1}{3}x^3$ is a vertical stretch of the graph of $y = -x^3$ by a factor of $\frac{1}{3}$

vertical translation of a graph another name for vertical shift; for example, the graph of $y = -\frac{1}{2}x^2 + 7$ is a vertical translation of the graph of $y = -\frac{1}{2}x^2$ up by 7 units

W

word problem popular math-class exercise in which a math problem is presented in a contrived real-world scenario

X

x-axis usually, the horizontal axis of a coordinate plane; its coordinates increase from left to right

x-coordinate usually, the first of two numbers in an ordered pair; for example, the 3 in (3, –7)

x-intercept the point where the graph of an equation crosses the *x*-axis; also, the *x*-coordinate of that point

xy-plane another name for the coordinate plane

Y

y-axis usually, the vertical axis of a coordinate plane; its coordinates increase from bottom to top

y-coordinate usually, the second number in an ordered pair; for example, the –7 in (3, –7)

y-intercept the point where the graph of an equation crosses the *y*-axis; also, the *y*-coordinate of that point; (0, *b*), or just *b*, for the equation $y = mx + b$

Z

zero of an expression a value for the variable that makes the expression evaluate to zero; for example, 4 is a zero of $2 - \sqrt{x}$; zeros of polynomial expressions are known as roots

zero polynomial the polynomial 0, which has no degree

LESSON

Notes

Notes

Notes

Notes

Notes

Notes

Notes

Notes